ECONOMIC ANALYSIS OF PROVINCIAL LAND USE POLICIES IN ONTARIO

Mark W. Frankena and David T. Scheffman

Economic Analysis of Provincial Land Use Policies in Ontario

PUBLISHED FOR THE ONTARIO ECONOMIC COUNCIL BY
UNIVERSITY OF TORONTO PRESS
TORONTO BUFFALO LONDON

Reprinted 2017
ISBN 978-0-8020-3364-2 (paper)

Canadian Cataloguing in Publication Data

Frankena, Mark W., 1943-
Economic analysis of provincial land use policies in Ontario

(Ontario Economic Council research studies ; 18
ISSN 0708-3688)

Bibliography: p.
ISBN 978-0-8020-3364-2 (pbk.)

1. Land use – Planning – Ontario. 2. Regional planning – Ontario. 3. Ontario – Economic policy. I. Scheffman, David T., 1943- II. Title. III. Series: Ontario Economic Council. Ontario Economic Council research studies ; 18.

HD319.05F73 333.73'17'09713 C79-094853-2

This study reflects the views of the authors and not necessarily those of the Ontario Economic Council.

Contents

Preface

This study presents an economic analysis of provincial land use policies in Ontario for use by a wide audience: students and professionals with backgrounds in economics, geography, or planning; government officials; and the general public. Our aim in carrying out the study has been twofold. First, we have tried to bring together in one place a rather considerable amount of material on the subject. Second, and more important, we have tried to explain the contribution that economic analysis can make to the intelligent formulation and evaluation of land use policies. In doing so we have examined the methodology which economists have developed to evaluate government intervention in the economy, and we have analysed the economic justifications for several provincial land use policies in Ontario.

Our hope is that this study will bring economists, planners, and policy-makers closer together on matters concerning land use. For economists, this study provides a survey of what the provincial government has been doing to influence the allocation of land. For planners and policy-makers, it explains how economic analysis ought to be used in the formulation of land use policies. Ultimately, we hope that the study will help to modify the traditional curriculum of planning education in Canada in which economics is almost totally neglected. Planning is, to a very large extent, applied economics, and this fact should be recognized both in schools of planning and in the government.

Of course we also hope that members of the general public who are neither economists nor planners will find the study an informative source of both description and critical evaluation of several land use issues which are of widespread concern. Anyone who is concerned with matters such as the conversion of farmland, or with other current issues raised by documents such as the reports of the Planning Act Review Committee and the Federal/Provincial Task Force on the Supply and Price of Serviced Residential Land and the province's *Green*

Paper on Planning for Agriculture, will find here useful background information for discussion of public policy.

Although all of the data and institutional descriptions refer specifically to Ontario, most of our analysis is also relevant to land use policies in any part of North America. Consequently, the volume should be useful to readers outside Ontario.

Many people have provided us with information and suggestions in the course of this study. George Fallis was very helpful in the process of determining the initial terms of reference. Ralph Krueger, Stephen Rodd, and Lorne Russwurm offered suggestions and made some of their unpublished work available to us. We received many useful suggestions from the participants in two review seminars organized by the Ontario Economic Council in July 1977 and June 1978. We also benefited from extremely valuable written comments on drafts of some of the chapters from John Bossons, Eli Comay, Len Gertler, and Brock Smith and on the penultimate draft of the entire report from Enid Slack and two anonymous referees. Of course, it should not be assumed that any of these people agree with our approach or conclusions.

ECONOMIC ANALYSIS OF PROVINCIAL LAND USE POLICIES IN ONTARIO

1

Introduction

As in most areas of economic activity, the post-war period has seen a marked increase in the role of all, and especially senior, levels of government in Canadian land and housing markets. Consistent with this general trend, there has been a widening of the powers of the provincial government in Ontario over the allocation of land. The driving force behind this development has been the growth and urbanization of the economy. For various reasons federal and provincial governments have felt that the response of municipal governments to the pressures of growth and urbanization has been inadequate or inappropriate.

Despite the fact that the centralization of political authority in urban affairs is common to most areas of North America, there has been a paucity of economic analysis of the justification for and suitability of such centralization. This is largely because urban economics, as an analytical discipline at least, is fairly new, and thus far most research on public policy in urban areas has been concerned with the role of local governments.

In the present study we shall describe and discuss provincial government land use policies in Ontario and use the tools of economic analysis to examine whether such provincial intervention is justified on economic grounds.

THE ROLE OF ECONOMICS IN POLICY FORMULATION

As economists, we are concerned with the *economic* justification for and the effects of government policies. Although we are not so naïve as to believe that economic evaluations either are or should be the only input to the political process by which policy decisions are made, we strongly believe that, for any proposal which would affect the allocation of scarce resources or the distribution of income and wealth, an explicit assessment of economic costs and benefits and their distributions is an indispensable input to the intelligent formulation of public policy. We shall argue in this study that such cost-benefit calculations have not been employed adequately in the past in the formulation or review of the provincial government's land use policies, so that at times the economic justification for provincial policies has remained open to doubt.

The issue here could, of course, be regarded as a matter of who bears the burden of proof, that is, of whether the government should be required to justify its intervention in the economy, or opponents of each proposed government intervention should be required to demonstrate that the intervention in question is not justified. Our position, quite simply, is that normally the government should be expected to carry out or sponsor an assessment of economic benefits and costs and their distributions and to publicize the results prior to committing itself to a course of action. This does *not* mean that government intervention should never take place unless it has been proved beyond a shadow of a doubt that the economic benefits would exceed the costs. Obviously, for example, there will be situations in which difficulties in measuring benefits and costs will leave a considerable range of uncertainty about what the appropriate government action is. However, the recognition that there are limits on cost-benefit analysis is not a justification for failing to consider economic benefits and costs as carefully as circumstances permit.

There are two rather distinct reasons, we would argue, why the onus should be on the government to justify intervention in the economy. First, there is ample evidence that governments at all levels have intervened in ways which have led to a less efficient allocation of resources and a less equitable distribution of income. This intervention may be explained, at least in large part, by the incentive politicians have to increase their prospects for re-election and by the ambition bureaucrats have to maximize the size and power of their agencies. In short, it would be beneficial to subject the government to the increased discipline and accountability that would be involved in a process of economic evaluation of proposed interventions. Second, as a practical matter, only the government has the resources to carry out or sponsor systematic economic evaluation of its policy proposals. The expenditure involved would, indeed, represent only a relatively small share of the government's budget, while being beyond the means of the organizations and individuals which now take an interest in monitoring the economic benefits and costs of government policies.

It is important to realize that our recommendation that the government should carry out economic evaluations of its proposed interventions would do nothing to bring about desirable government intervention if the government was not motivated to intervene. For example, if the government had no interest in creating an efficient system of parks or in taking efficient measures to limit environmental degradation, our recommendation would have no effect. This is a serious problem, but obviously it would be unrealistic to expect the government to undertake economic evaluations of every potential intervention. It would, however, be in the spirit of our recommendation to have the government adopt a policy of undertaking or sponsoring, on a larger scale than it currently does, periodic economic evaluations of certain issues (such as the adequacy of the parks system, of environmental protection policies, or of land use policies in northern Ontario) to determine whether additional government intervention would be efficient.

AN OUTLINE OF THE STUDY

We begin in Chapter 2 with an explanation of the methodology of economic policy analysis, and then apply this methodology in a general way to analyse the economic case for provincial intervention in land markets. We explain that an economic case for intervention can be made on efficiency grounds if market imperfections have resulted in a market failure, i.e., the inability of markets to allocate resources efficiently. The major potential sources of market failure (externalities, public goods, uncertainty, etc.) in this context are critically evaluated. We also discuss the economic case for intervention on grounds of equity and the role of distributional effects in economic policy evaluation.

Chapters 3 to 7 provide a description of provincial land use policies in Ontario and an economic analysis of the province's intervention in land markets.

To begin, Chapter 3 provides a general overview of the many areas of provincial land use policy in Ontario. We first survey provincial control of local planning under the Planning Act and recent proposals to reform the provincial supervisory role emanating from the Comay and Robarts reports in 1977, and then describe several other important areas of provincial land use policy: provincial regional planning under the Design for Development program; direct provincial planning under the Ontario Planning and Development Act, in the Parkway Belt West, and along the Niagara Escarpment; provincial planning and land acquisition for new towns; and the numerous activities of the Ministry of Natural Resources and the Ministry of the Environment in the area of land use controls. We also point out that many government policies other than explicit land use controls – e.g., agricultural and housing subsidies and provision and subsidization of interurban transportation and major sewage facilities – have important effects on land use in the province.

Based on this general survey of provincial intervention in land markets, we have selected three important areas of provincial land use policy for more detailed examination and evaluation in succeeding chapters: (1) the provincial role in the municipal planning process (Chapter 4), (2) provincial policies regarding rural and agricultural land (Chapters 5 and 6), and (3) provincial regional planning (Chapter 7). In each case we have attempted to provide a careful description of provincial policies and to assess the economic justifications for them.

In Chapter 4 we examine whether there is an economic case for provincial intervention in the municipal planning process on the grounds that municipal governments may enact policies which reduce the efficiency of local resource allocation. Our approach has been to develop an economic theory of development controls, elaborated in Frankena and Scheffman (1979) and summarized in this chapter. This theory identifies the major economic incentives for municipalities to control development and explains what sort of policies will arise from these incentives. The analysis provides a framework within which one can examine the question of whether municipalities will enact policies which reduce efficiency. In particular, the question of whether municipalities have market power and are likely to exercise it is considered in detail. Finally, the recommendations of the Planning Act Review Committee are critically evaluated.

When we reviewed the debate over rural and agricultural land use policy, we discovered that there was considerable factual ignorance on the part of concerned parties regarding present patterns and recent trends of land use in rural areas. Because an understanding of these matters is necessary for a sensible evaluation of policy, Chapter 5 brings together in one place for the first time the available data on two key aspects of land use: (1) rural non-farm residences and (2) conversion of agricultural land to other uses.

The data assembled in Chapter 5 demonstrate clearly that in the aggregate the rate of conversion of agricultural land to built-up urban use and non-farm rural residential use in Ontario has been low in relation to the rate of productivity increase in agriculture, the stock of agricultural land, and the decline in the acreage of census farms. Although these findings do not imply that public policy should not be concerned with the rate of agricultural land conversion, they do make it apparent that there is little factual basis for the alarmist rhetoric which has sometimes characterized recent discussions of the agricultural land issue. We believe that anyone interested in the farmland issue would do well to study carefully this chapter and the following one.

Based on the foundation laid in Chapter 5, in Chapter 6 we survey provincial policies concerning rural and agricultural land. We begin with a review of the provincial 'Urban Development in Rural Areas' policies which since 1966 have been aimed at restricting non-farm residential development in rural areas. We then review the more recent policy of preserving prime and unique agricultural lands for agricultural use, including the province's 1977 Green Paper on Planning for Agriculture, and consider the Niagara urban area boundaries issue, which was the principal test of the province's new policy toward agricultural land. The chapter also includes an economic analysis of the case for provincial intervention in rural land markets and of the particular forms that provincial intervention has taken.

Several parts of Chapter 6 are critical of the government's policies, for two different reasons. First, we argue that the government has failed to carry out even the most rudimentary evaluation of the economic benefits and costs of its intervention in the allocation of rural and agricultural land. Second, we argue that the particular forms of provincial intervention have sometimes been inefficient.

Chapter 7 provides a review of provincial involvement in comprehensive regional land use planning under the Design for Development program, with particular attention to the Toronto-Centred Region Plan. We consider the evolution of this program over the decade and a half since its origins in the Metropolitan Toronto and Region Transportation Study and the factors which led to its virtual abandonment by the mid-1970s. In addition to analysing the economic justification for provincial regional planning, we review the evidence concerning the effect of the Toronto-Centred Region Plan on land prices. Our findings here are consistent with the 1978 report of the Federal/Provincial Task Force on the Supply and Price of Serviced Residential Land in Canada, which concluded that changes in demand conditions rather than changes in supply conditions (including government land use controls) provide the explanation for the escalation in land prices in the early 1970s.

As in Chapter 6, several parts of Chapter 7 are critical of the government's approach to land use controls, chiefly because the planning was carried out without systematic concern for economic benefits and costs or their distributions. Since the government never established the justification for its regional planning initiatives, we find no cause to regret the province's retreat from this program.

In Chapter 8 we summarize the findings presented in the preceding chapters and provide recommendations for changes in the government's approach to land use policy formulation.

2

The economics of provincial intervention in land markets

In this chapter we consider the possible economic justifications for provincial intervention in land markets, and develop an analytical framework for our evaluation of specific provincial policies in later chapters.

We should state at the outset that we realize that economic arguments are not the only determinant of policy decisions. However, economic analysis is indispensable to rational policy-making whenever the allocation of scarce resources or the distribution of income and wealth is affected. In the past economic analysis has not been employed sufficiently as an input in provincial land use planning or provincial decision-making with respect to land use policy. We hope that our analysis will prove helpful in developing a framework for making future land use policy decisions.

THE METHODOLOGY OF ECONOMICS

Economics is concerned with the allocation of scarce resources and the distribution of income and wealth in society. Economic analysis can and should be applied to any policy which has a significant effect on either of these factors. Thus, economic analysis should be applied to land use policies. Indeed, properly carried out, planning is to a large extent applied economics.

Perhaps the most important implication of economic analysis for policy-making is that in a world of scarce resources *trade-offs* characterize any policy decision. Non-economists often argue as though goals should be pursued relentlessly: all good agricultural land should be preserved for agriculture, all urban and residential development should be laid out in the manner which minimizes the cost of public services, pollution should be eliminated, the price of housing should be minimized, etc. This type of reasoning ignores the fact that pursuit of one goal often requires sacrifice of another.

Economists deal with trade-offs explicitly. Given the limited resources in the economy and the available technology, a variety of alternative *consumption bundles* are open to the people in the economy during any time period (and over time). Assuming for simplicity that there are only two consumption goods (food and housing) which can be produced with the available resources, the people in the economy could have a consumption bundle with mainly food and little housing, or one with mainly housing and little food, or various intermediate combinations. Each additional unit of food consumed requires the sacrifice of some housing which could have been consumed instead, i.e., the *opportunity cost* of additional food in terms of forgone housing is explicitly recognized.

The important problem facing policy-makers is to choose among alternative consumption bundles and their distributions that would result from adopting different public policies. The approach of non-economists offers no systematic framework for evaluating such choices. Economists normally evaluate these choices in terms of (1) the *efficiency* of resource allocation and (2) *equity* or the distribution of income.

A policy is efficient if the aggregate benefits accruing to all members of society exceed the aggregate costs and the excess of benefits over costs is as great as that of any mutually exclusive policy alternative. Benefits from land use policies typically include such things as increased output of certain goods and services (e.g., recreational services, food), reduced input requirements for production of public services, and reduced pollution damage. These benefits are valued in money terms at the amount that the people affected by the policy would be willing to pay for them. Costs, or opportunity costs, are the things that people must forgo as a result of the policy. They typically include such things as forgone output of other goods and services (e.g., housing), increased input requirements for transportation, and inputs used by the government to administer the policy. The costs can be evaluated at the amount of money that the people affected would have to receive if they were to be compensated for bearing them.

Thus, not only does the efficiency criterion make it clear that both benefits and costs must be considered in evaluating policy alternatives, but most important of all this criterion provides a basis for measuring benefits and costs on a common scale, namely willingness to pay in dollars on the part of all people affected.

However, one cannot choose among policies on the basis of *aggregate* benefits and costs, or efficiency, alone. It is typical that a policy which will show an excess of benefits over costs in the aggregate will still make some groups in the population worse off. Economists sometimes ignore such issues of equity or income distribution on the grounds that, if a policy is efficient, the government could redistribute income from the gainers to the losers in such a way that everyone could be left better off. However, most economists now recognize that this argument does not justify neglect of distributional effects, for three reasons. First, despite what intuition might suggest, the existence of positive aggregate net benefits is neither necessary nor sufficient for compensation of losers by gainers to be possible, even if such redistribution were costless.[1] Second, such a redistribution would in fact have costs in terms of resources used for administration and distortion of resource allocation caused by the taxes and subsidies used to carry it out. Third, in practice governments simply do not carry out such redistributions to compensate losers. Any sensible evaluation of public policy must therefore consider not simply the (unweighted) total value of net benefits. Instead, it must consider the distribution of the net benefits and net losses among the members of the population, the relative weights to be placed on welfare changes for different groups, and the *weighted* sum of net benefits and net losses across all groups. This type of evaluation is what we have in mind when we advocate economic evaluation or cost-benefit analysis of public policy in this study, even though we do not always call attention explicitly to distributional effects.[2]

Adam Smith's Invisible Hand Theorem

One of the most important contributions of modern economic analysis has been a proof of what might be called 'Adam Smith's Invisible Hand Theorem.' This theorem states that in the absence of what economists refer to as 'market imperfections' (which will be described below), free markets will allocate society's resources efficiently. The concept of efficiency used here is what economists term *Pareto efficiency*. An allocation of resources is efficient in this sense if it cannot be changed so as to benefit any individual without decreasing the well-being of another. Alternatively, if an allocation is not efficient, it is possible to change it so that some or all individuals are better off and none are worse off. Full (Pareto) efficiency requires the equality of marginal *social* costs of production and prices. This condition ensures that the full social costs of production are minimized, and that prices correctly reflect social costs.[3]

It is important to recognize that even when a market allocation is efficient, it may still be regarded as inequitable, and government intervention in the economy may be considered desirable in order to change the distribution of income.

1 See Boadway (1974). The compensation itself would normally change relative prices.
2 For discussions of how distributional effects might be integrated into cost-benefit analysis, see Sugden and Williams (1978) and Boadway (1974).
3 For a more thorough development of the concept of Pareto efficiency, see Hirshleifer (1976) or Orr (1976).

In this event, economists usually argue that the redistribution should be carried out by means of a policy which has the least possible adverse effect on the efficiency of resource allocation, such as by a cash transfer. Clearly, if one is interested in effecting a particular redistribution of income, and if an explicit cash transfer is feasible, it would not be sensible to undertake an inefficient land use policy which would distort resource allocation and prices simply in order to achieve the redistribution.

The standard economic methodology used to evaluate any proposed government intervention in the allocation of resources starts with Adam Smith's theorem. The theorem states that, in the absence of imperfections, markets will allocate resources efficiently. Imperfections in this context are deviations from an idealized frictionless world of atomistic economic agents, termed the model of perfect competition by economists. Consequently, the economist begins by determining whether such imperfections exist. The failure of a market to allocate resources efficiently is termed a *market failure*, and such imperfections are therefore *potential sources of market failure.*

Showing the existence of a potential source of market failure does not, in itself, make an argument for government intervention because, as we shall see below, there are often market or non-market institutional arrangements which satisfactorily ameliorate potential sources of market failure.[4] Furthermore, intervention is not costless, and the direct and indirect costs of implementing any proposed policy must be considered. The economist's final step in assessing a proposal for government intervention must therefore be an examination of the institutional arrangements specific to the problem at hand, and an assessment of the administrative and other costs of intervention.

The importance of this final step cannot be overemphasized. The urban planning literature has become increasingly familiar with economic jargon in recent years, and hence the arguments for any proposed planning action generally include references to potential sources of market failure. (For example, the obvious existence of externalities present in location decisions is given as a prima facie case for the necessity of planning for the efficient allocation of resources in an urban area.) Such arguments, however, are invalid applications of economic methodology if they do not include consideration of any relevant non-market institutional arrangements and an assessment of the costs of intervention.

One of the more striking pieces of evidence which casts doubt on the validity of planners' typical invocation of potential sources of market failure as a justification for zoning and planning controls is the example of Houston, Texas, where

4 For an excellent discussion of the institutions which have evolved in the medical care industry to deal with potential sources of market failure, see Arrow (1963).

government zoning controls are not used and control of land use is exercised through *private* covenants in deeds.[5] A final verdict on the success of the Houston system cannot be made at this point because of an insufficient body of evidence and analysis, but the example indicates the possible importance of institutional arrangements in dealing with potential sources of market failure.

Intertemporal efficiency and equity

Many of the more important potential sources of market failure in land and housing markets arise in an intertemporal context. For example, the unknown course of future events is obviously the major source of uncertainty faced by economic agents. The classical version of Adam Smith's theorem assumes that markets exist for all relevant commodities, which in an intertemporal context would require that there be a complete set of futures markets for all commodities. Because in the real world markets cannot be created and maintained costlessly, and agents must bear transactions costs, the private economy does not find it profitable to have such a comprehensive set of markets. However, the private economy is especially ingenious in creating substitutes for these uneconomical markets. Capital markets and legal contracts, for example, are just two vehicles by which an agent can engage in intertemporal trades. Nonetheless, potential sources of market failure may impinge on the ability of the private economy to allocate resources efficiently over time, and this issue will be fully discussed below.

The problem of equity gains an additional dimension in an intertemporal context, because many of the agents (future generations) affected by current market choices have no direct voice in those choices. Thus decisions of the current generation which increase or decrease the capital stock, the stock of agricultural land, etc. obviously have an important impact on the well-being of future generations. In the terminology of simple economic models, economists describe the intertemporal allocation of resources by the private economy as being governed by the *private discount rate*, which is the private market economy's intertemporal terms-of-trade ('the' interest rate serves this function in simple models). If the private market economy undervalues the well-being of future generations relative to society's criterion of fairness, the situation is described as one in which the private discount rate is larger than the social discount rate. In such a situation the private market economy excessively discounts the value of investments which take a long time to pay off and underinvests (relative to society's criterion of optimality). This is a problem of particular potential importance in the context of land markets, where it is often

5 Siegan (1972).

argued, for example, that since the desired (by the arguer) social discount rate is below the private discount rate, agricultural land is converted to urban use at an excessive rate. The argument here is that an expected future food scarcity will eventually make agricultural land much more valuable, but the current excessive private discount rate discounts this future value too heavily.

As an example, consider a stationary economy with a zero inflation rate in which an acre of land has an implicit rental value of $10,000 a year in perpetuity if used for housing. With a private discount rate of 10 per cent, this land is worth $100,000 if used for housing. Suppose further that the current annual agricultural rent is $1,000 per acre, but it is known that the annual agricultural rent will rise to $50,000 an acre in 20 years and remain there in perpetuity. With a discount rate of 10 per cent, the land is worth about $40,000 if used in agriculture. In this situation, of course, the land will be converted to urban use. However, if the social discount rate is 5 per cent, the value to society of an acre of land used in housing is $200,000, but the value to society of an acre of land used in agriculture is almost $400,000. Therefore, from society's point of view, the land should be kept in agricultural use (assuming it is sufficiently costly to reconvert land used for housing back to agricultural use).

Thus, a concern with intergenerational equity may itself justify government intervention in land and housing markets, but this argument imposes considerable informational requirements: the social discount rate, the value of land in different uses at different times, the cost of reconversion, etc.

Cost-benefit analysis

There are excellent texts intended for non-economists which explain and illustrate cost-benefit analysis, so we shall not provide a detailed discussion of how such an analysis should be carried out.[6] Nevertheless, because we are recommending that cost-benefit analysis be used much more extensively in the formulation and evaluation of provincial government policy, we might outline what is involved in the typical analysis of a government project or policy. (In addition, in Chapter 6 (pp. 113-14), we shall provide an example of how a cost-benefit analysis of the policy of preserving prime agricultural land might be carried out.)

The first step in any cost-benefit analysis is identification of the benefits and costs of the project or policy. The benefits typically involve additional production and consumption of various goods and services, or additional supplies of scarce resources which can be used to produce additional goods and services. The

6 For a lucid text on cost-benefit analysis, see Sugden and Williams (1978). For a variety of empirical case studies, see Dorfman (1965), Wolfe (1972), and Frankena (1979), Chapter 7.

costs typically involve reduced production and consumption of other goods and services, or use of scarce resources which could otherwise have been used to produce other goods and services.

For example, consider a pollution abatement policy which induces a pulp and paper factory (a) to invest in a pollution control device which reduces its emission, per ton of paper, of pollutants into a lake and (b) to reduce the level of output of paper. Suppose that the reduction in pollution leads to an increase in the number of eatable fish caught by commercial fishermen on the lake, an increase in the quality of the lake for recreational purposes, and a reduction in the costs of treating lake water to make it fit for use in a nearby city.

In this example, the benefits of the policy are (1) the increase in consumption of fish (net of any increase in scarce resources devoted to fishing), (2) the increase in consumption of recreational services (net of any increase in scarce resources devoted to recreation), and (3) the reduction in scarce resources used to treat lake water used by the city. The costs of the policy are (1) the use of scarce resources for the pollution control device and (2) the reduction in consumption of paper (net of any reduction in scarce resources devoted to paper production).

As a second example, consider a policy which would create a park in an urban area. The benefits here would probably include (1) an increase in consumption of recreational services, such as cross-country skiing, and (2) a reduction in resources devoted to recreational travel, such as to rural ski areas. The costs might include (1) a reduction in consumption of urban residential land and food and (2) an increase in resources devoted to urban commuting, since allocation of land to the park would lead to a reduction in lot sizes and an increase in the area of the city. Alternatively, of course, consumption of food might be maintained by devoting more non-land resources to production of food, but in this case consumption of some other good would decrease.

It is important to point out that in identifying effects which are in fact benefits or costs in terms of the efficiency criterion, one is concerned only with effects which involve changes in *aggregate consumption* of goods and services for all people taken together. A project or policy may have other effects (for example, changes in land prices or tax liabilities) which involve transfer payments which affect only the distribution of income and wealth, and these should not be confused with aggregate consumption or efficiency effects. Such transfer payments are relevant to the evaluation of the equity of a policy (which is discussed further below), but to avoid confusion such distributional effects should be evaluated separately from the aggregate benefits or costs measured in terms of the efficiency criterion.

The second step in any cost-benefit analysis is to value the benefits and costs in dollar terms. According to the efficiency criterion, each benefit should be assigned a dollar value equal to the amount that all people affected taken together are willing to pay for it. Similarly, each cost should be assigned a dollar value equal to the amount that all people affected taken together would have to be paid to be compensated for bearing it.

Although it is often difficult to place a dollar value on some of the benefits and costs of a policy, three points should be emphasized. First, there is already a very extensive empirical literature dealing with difficult problems of valuation, such as valuation of recreational facilities, valuation of improved health, and valuation of reductions in congestion and pollution. If the government would use cost-benefit analysis more extensively, further progress in the valuation of various types of benefits and costs would be rapid. Second, often it turns out that a cost-benefit calculation does not in fact require a valuation of every effect–for example, it does not matter that one cost item cannot be valued when the sum of the other cost items already exceeds the total benefits. Third, even if it proves impossible to assign a dollar value based on willingness to pay for some category of benefits or costs, cost-benefit analysis provides a useful framework for determining trade-offs. For example, suppose that one category of benefits cannot be valued in dollar terms. A cost-benefit analysis will still make it possible to determine the net cost (all costs minus all other benefits) required to obtain this particular non-quantifiable benefit, and a rational political decision on the policy can then be made on the basis of an explicit consideration of the trade-off between the non-quantifiable benefit in question and the net cost in terms of aggregate consumption of other goods and services. Thus, even if certain costs and benefits are non-quantifiable, cost-benefit analysis provides a useful framework for sensible discussion of policy alternatives.[7]

The third step in any cost-benefit analysis is using the analysis to reach conclusions concerning policy. At this point it is important to keep in mind that there may be legitimate political objectives in establishing policy other than increasing the efficiency of resource allocation, such as increasing the fairness of the distribution of income and wealth or reducing economic disparities among regions. Since a cost-benefit analysis will quantify the efficiency (i.e., aggregate consumption) gains or losses from a policy, it will make it possible to consider explicitly the trade-off between efficiency or aggregate consumption and other objectives. An explicit consideration of the aggregate consumption costs of

7 There are, of course, many other difficulties in applying cost-benefit analysis, such as uncertainty about the social rate of discount and problems in determining general equilibrium effects.

achieving other objectives, such as reduction in economic disparities, is indispensable for rational decision-making in a world of scarce resources. Of course, in some cases efficiency and other objectives will not conflict.

Absence of economic evaluation of provincial land use policies

An economist who reads the provincial documents on land use policy in Ontario will be struck by two basic weaknesses in their economic reasoning. First, they do not provide a coherent theoretical or empirical justification for provincial government intervention in the market for land based on specific market failures. Second, they do not contain careful consideration of opportunity costs of proposed policies, let alone systematic cost-benefit calculations. These two points are discussed in more detail below. As an example of the lack of careful justification for provincial land use policies, we shall consider the province's rationale for its regional planning program and the Toronto-Centred Region (TCR) Plan. (The province's regional planning program and TCR plan will be discussed fully in Chapter 7)

(a) Rationale for provincial land use policies

The basic motivation behind the intervention of the province in regional land use planning was essentially dissatisfaction with certain patterns which seemed to be evolving during the period of rapid population growth in the post-war period, such as concentration of growth in the largest urban areas and in the central and southwestern parts of the province, urban sprawl and urban development on prime agricultural land, and pollution of the environment.[8]

The general rationale offered by the provincial government and its various task forces for the Design for Development regional planning program and the TCR plan was that private markets and municipal government planning would not lead to the socially optimal allocation of resources. The following two quotations from provincial planning documents provide the most explicit statements of this rationale that could be found:

Urban development is always planned, as are most human activities. The point is: who is doing the planning and for whom is it being done. In the case of a trends plan (i.e., in the absence of government planning), it is every man for himself. Individual people and companies plan for their individual benefit without much regard for the sum of their efforts. Although it is possible that the results achieved will be to everyone's satisfaction, the chances of this occurring

8 Thoman (1971), pp. 29–32.

are slight. The more usual result is that everyone's plans are frustrated and no one is happy with the outcome.[9]

Change has brought a growing awareness of problems that must be dealt with, issues that must be resolved, at the level of the whole province. The days are gone when each locality was largely self-sufficient and what happened there was of little importance elsewhere. Today, the consequences of changing demographic, economic, and urban patterns transcend local and even regional boundaries. For example, economic difficulties in northern and eastern Ontario are at least in part related to the growing concentration of people and production in the south-central part of the province. The prosperity of individual communities throughout Ontario, and the services they are able to provide both to their own citizens and to surrounding areas, are increasingly dependent on their place in the urban system of the province as a whole. Steady encroachments on the natural resources of the province - far from unlimited, as they were once assumed to be - are a matter of concern for every Ontarian, whether he lives in Toronto, Trenton, or Timmins. The message is clear. The Government of Ontario has a responsibility to the people of the province to maintain and enhance the environment in which they live and the quality of life which they enjoy. To help them realize their aspirations, the government must work towards the optimum development of the province's economic potential, the wise use and protection of its resources and the equitable distribution of opportunity across the province.[10]

The first of these quotations is particularly curious from the economist's point of view. As explained earlier, it conflicts with 'Adam Smith's Theorem,' which states that in the absence of certain imperfections the decentralized market system will lead to an efficient allocation of resources. The case for planning rests heavily on establishing the empirical importance of market failures, something which is typically ignored and never done. There appears to be a widespread belief among non-economists that it is self-evident that decisions concerning the location of economic activity and the allocation of land are too important to be left to market forces, and that a superior allocation of resources will result if the government intervenes and allocates land according to the principles of conventional planning wisdom, which enunciate that, for example, our largest cities are too big, urban sprawl must be prevented, or the best

9 Ontario, Central Ontario Lakeshore Urban Complex Task Force (1974), p. 1.
10 Ontario, Ministry of Treasury, Economics and Intergovernmental Affairs, Regional Planning Branch (1976a), pp. 1–2.

agricultural land must be preserved for agriculture. Such principles are irrational because they neglect the opportunity costs of the policies they support. Yet, the belief in them is evidently so general that non-economists often do not feel the need to justify government intervention with in-depth analysis.

Of course, our argument is not intended to suggest that there are no market failures which would justify existing or other possible provincial land use controls. We are simply suggesting that the province often has not bothered to establish an economic justification for its land use policies, and that in many cases it remains an open question whether such a justification could be provided.

(b) Neglect of cost-benefit considerations

The TCR planning documents do not go beyond the very general statements quoted above to provide a cogent justification for specific types of government intervention to correct specific market failures due, for example, to the existence of externalities or public goods. Instead, the provincial planners justify the plans on the grounds that they would move the economy toward certain goals, such as minimizing the costs of transportation and waste disposal, conserving land to meet future demands for recreation, and reducing regional economic disparities. Although most of the goals proposed appear acceptable, there is a critical gap in the argument. The planners may be right that their plans would have benefits of the type alleged, but they almost invariably ignore the unavoidable opportunity costs of moving toward these goals, such as the reduction in housing consumption which would result if conversion of land from agricultural to residential use was restricted. In short, moving toward one goal (more food) generally involves moving away from another goal (more housing). The planners have generally failed to make a convincing case that the planned position would be preferable to what the market would produce. This statement does not, of course, imply that the planners are necessarily wrong, but it is evident that the planners and politicians did not systematically consider the costs as well as the benefits of regional planning when the TCR plan was being formulated and adopted in 1969-71.

The absence of cost-benefit considerations in provincial planning has not gone unnoticed by others. For example, the report of the Parkway Belt West Interested Groups and Residents Advisory Committee states: 'We are aware of the goals and objectives as listed in the Plan and find these desirable ... However, nowhere do we find any assessment or statement of the costs either to the public at large or to these landowners directly affected by the introduction of the Plan. Without information on costs which can be considered together with a listing of the benefits expected, it is difficult for persons considering the Plan to come to informed judgements on net benefits or costs of the Plan.'[11]

The neglect of opportunity costs in land use planning is not unique to the TCR plan. According to the Advisory Task Force on Housing Policy (Comay) Report, prior to 1973 the forgone consumption of housing services was not considered when residential development was subjected to an increasing number of planning controls in Ontario:

> Many governmental activities affect housing, at both the provincial and municipal levels. In carrying out these activities little attention is paid to their effect on the supply or the cost of housing. Regulations and administrative procedures are protective and negative with regard to housing, rather than positive and productive. Servicing and development standards are set and applied without reference to their effect on housing. Departmental programs are pursued without regard to their implications on housing supply or cost ... In planning and development, at the local, regional and provincial levels, housing has not received conscious consideration.[12]

Similarly, recent provincial statements concerning the preservation of agricultural land essentially ignore the opportunity costs of preserving land for agriculture. These statements, like much of the public discussion of the issue, regard the proposition that the best agricultural land should be preserved for agriculture as axiomatic and beyond the need for justification based on an analysis of benefits and opportunity costs.

POTENTIAL SOURCES OF MARKET FAILURE

We shall now assess the potential sources of market failure which, in the absence of suitable institutional arrangements, may justify intervention by the *provincial* government in the allocation of land. There is an extensive literature which discusses the possible role of *local* governments in land markets. Since the present study is concerned with the provincial government's involvement in land markets, we shall consider the usual arguments for local government intervention only in so far as they relate to provincial policies. (One legitimate provincial concern may be that local authorities do not always adopt the appropriate policies for dealing with local problems, and this issue will, of course, be

11 Ontario, Parkway Belt West Interested Groups and Residents Advisory Committee (1975), pp. 2–3.
12 Ontario, Advisory Task Force on Housing Policy (1973c), pp. 38–40.

examined in detail later, in Chapter 4.) We discuss local zoning, for example, only to the extent that it bears on provincial policy (which should include a concern with the equity and efficiency of local zoning policies) or the interurban efficiency of resource allocation.

Market power

An economic agent is said to have market power if he can exert significant control over the price and/or output in a market. When an agent has market power, it will generally be in his interest to exercise that power, and if he does a market failure will result. In the simple textbook case of monopoly (a market with a single seller), it is in the monopolist's interest to restrict output below the efficient level, which raises the price above social marginal cost.

A thorough examination of the determinants of potential market power and the allocative effects of exercised market power can be found in Markusen and Scheffman (1978a; 1977a, b). Furthermore, documentation of concentration of land ownership and housing production in the land and housing markets in Toronto by Markusen and Scheffman (1977a, b) and Muller (1978), in London by Whitney (1977), and in several Canadian cities in the report of the Federal/ Provincial Task Force on the Supply and Price of Serviced Residential Land (1978) indicates that the development industry probably does not possess market power in most urban land and housing markets in Ontario. It is therefore not necessary here to provide a detailed discussion of market power in land and housing markets.

Another potential source of market failure is the possible exercise of market power by local governments through their control of development. This issue was given prominence in the report of the Federal/Provincial Task Force on the Supply and Price of Serviced Residential Land, and this type of market power suggests a possible role for provincial intervention. Although the exercise of market power by local governments has received some attention in the economics, legal, and geography literature, there is no thorough analysis of the determinents of when such power will exist, when it will be exercised, the allocative effects of its exercise, and the possible role for provincial intervention. We shall develop such an analysis in Chapter 4. Suffice it to say here that our analysis shows that growing urban areas will generally find it in their interest to exercise market power, but that such policies are not necessarily inefficient. Furthermore, the scope for enacting provincial policies which would increase efficiency is small.

Finally, we might note that perhaps the most important agent possessing market power in land and housing markets is the provincial government itself, through its control of the subdivision approval process and its role in the provision of major servicing and transportation networks, a point that will be argued in Chapter 4.

Externalities

An externality exists when the economic actions of one agent (consumer or producer) have a non-market effect on another. For example, if a fossil-fuel electrical generating plant is located near me, I will be forced to 'consume' some of the smoke and dust it produces. This smoke and dust would be termed a negative externality. Externalities (both positive and negative) are possible sources of market failure because they may cause deviations between marginal social costs and prices. In the power plant example, the social costs of generating electricity include the damages imposed on me and other affected parties. However, in the absence of suitable institutional arrangements or government intervention, the managers of the generating plant will consider only private costs in their production decisions, which would result in a market failure. In this case, both the level of emission per unit of electricity and the level of electricity generated would exceed the efficient levels.

One of the most important institutions used for the non-market allocation of externalities is the legal system.[13] If, for example, the affected parties can successfully sue the electrical generating plant for the damages incurred in the situation described above, and if the transactions costs of suing are not excessive and the legal system properly assesses damages, the managers of the generating plant will be forced to recognize the full social costs of production.

Externalities are usefully divided into two general categories: small-number and large-number externalities, where small or large refers to the number of agents involved with the externality. In an idealized world with no transactions costs and where all agents have complete information about the effects of their and other agents' actions, we would expect that the existence of externalities would not result in market failures. The basis of this argument is that any deviation from efficiency, by definition, means that there is a potential gain for all agents involved by increasing efficiency, because the gainers could compensate the losers and leave everyone better off. In the 'real' world, transactions costs and informational imperfections will generally be directly related to the number of agents involved. Large-number externalities are therefore likely to be potential sources of market failure which private institutional arrangements cannot ameliorate. In such situations there is a potential case for intervention by

13 For an economic analysis of the legal system, see Posner (1977) and Manne (1975).

some level of government. Of course to make such a case it must be argued that 'the' government can sufficiently economize on private transactions costs or the private channels for transmission of information.

The common sources of possible externalities in land and housing markets are location decisions, pollution, and congestion. Externalities local to a particular municipality are best dealt with (if at all) by local government action, such as zoning ordinances, or the legal system. The most likely cases for provincial intervention are macro-urban externalities, interurban externalities, and agricultural-urban externalities, which we shall now consider.

(a) Macro-urban externalities

There is a substantial economics literature dealing with 'optimal city size.'[14] This is concerned with the issue of whether urban areas will grow to or be limited to their 'optimal' size without government intervention. Most of this literature is based on the assumption that urban areas are characterized by agglomeration economies or diseconomies which are external to producers or consumers. For example, larger urban areas have larger and perhaps more diversified labour forces which may allow producers to operate more efficiently through matching of available skills with requirements, and reduced turnover costs. In contrast, larger cities may be characterized by increased per capita waste disposal and transportation costs because of pollution and congestion. A market failure may result from such externalities if current and potential urban residents and producers respond to their *average* net benefits rather than the efficient signal, which is *marginal* social net benefits. (This issue will be considered again in Chapter 4.)

The existence of macro-urban externalities seems obvious, but the case for provincial intervention on these grounds is at best uncertain. On theoretical grounds, the obvious presence of both positive and negative externalities makes the relationship between the efficient and equilibrium city sizes ambiguous. This ambiguity has unfortunately not been resolved by the limited amount of empirical analysis addressed to the issue.[15] Also, because of the effects of growth on incomes and taxes, local governments are likely to intervene through their zoning and planning powers and industrial and commercial development strategies, and such intervention may result in efficiency. (A complete analysis of such intervention is developed in Chapter 4.)

14 For a convenient summary, see Richardson (1973). For a more advanced treatment of the issues, see Henderson (1977) and Kanemoto (1979).

15 See Richardson (1973).

Therefore a general economic case for provincial control of city size cannot be made. It certainly cannot be argued convincingly on economic grounds that large cities are always too large and that optimality requires moderate-sized cities, although these arguments are common in the planning literature and have appeared in some of the major planning documents in Ontario. Nonetheless, the province plays a major role in the determination of city size through its provision and financing of major servicing projects, roads, etc. The implications of this fact will be discussed below.

(b) Interurban externalities

On theoretical grounds, the existence of interurban externalities would seem to provide one of the stronger cases for provincial government intervention. The major possible sources of such externalities are migration, pollution, and congestion. (One particular source of concern regarding migration, that the development control policies engaged in by local governments may result in inefficient migration patterns, is an issue which will be considered in Chapter 4.)

Interurban, interprovincial, and international pollution are externalities which have received increasing attention and action in recent years. The industrial, commercial, or residential pollution of several bodies of water and the air in some areas of Ontario is of a significantly non-local nature, and government action (if required) would have to be above the local level. Since such environmental issues have been addressed elsewhere,[16] we shall not discuss them further here.

Finally, the provincial function in the provision of interurban transportation, water, and sewage networks has significant effects on the level of traffic, congestion, and air pollution experienced in various cities. For example, although in principle the local government can allocate traffic efficiently (at least in a second-best sense) given the number of commuters, the province has a very significant effect on the number of commuters through its provision and control of the interurban transportation network. We shall discuss the implications of this fact below.

(c) Agricultural-urban externalities

It has been argued by geographers and agricultural economists that rural development is a very important source of externalities.[17] Non-farm residences in rural areas may impose negative externalities if the residents fail to control weeds, use septic tanks which pollute the water, and congest rural highways. Agricultural

16 See Dewees, Everson, and Sims (1975) and Baumol and Oates (1975).

17 See Rodd (1976a) and Rodd and Van Vuuren (1975).

activities produce noise, odours, dust, chemical contamination, etc., which non-farm residences find unpleasant. These externalities will be considered in more detail in Chapter 6. We suggest there that many of these externalities can be dealt with by government intervention (e.g., local weed control ordinances) which does not involve a prohibition of rural non-farm residential development. Of more serious concern, therefore, is how the decision to permit rural development is made. Scattered rural development *may* represent a market failure, but this is an issue which can only be resolved empirically. On theoretical grounds, it might be argued that such a market failure, if it exists, may be in part the result of development control policies themselves. Control of development rights makes whatever land is approved for development more valuable, and therefore creates incentives for inefficient development. This is not to say that development control policies are inefficient, but rather that such policies may require additional instruments to ensure efficiency.

Public goods

A *pure* public good is a good (or service) which has the property that one agent's consumption of it does not affect the ability of other agents to consume it. The classic example of a pure public good is national defence. It is generally argued that pure public goods are sources of market failure, because private markets will be unable to supply them efficiently as a result of the 'free rider problem.'

To clarify the nature of the free rider problem, consider the following simple example. Suppose an economy with 100 identical consumers has one private good, denoted X, and one pure public good, denoted G. Suppose that each unit of G is produced at a cost of 100 units of X. Because the consumers are identical, each consumer would pay one unit of X for each unit of G supplied. Efficiency requires that G be supplied at a level at which each of the identical consumers would find his well-being unchanged by a one unit reduction in G combined with a one unit increase in X (for each consumer). (See Orr, 1976.)

The nature of the free rider problem is revealed by now considering whether the efficient provision of the public good could be sustained by free market forces. Starting at the efficient allocation, if any one individual reduced his expenditure on G by one unit of X, the level of G provided would fall by only 1/100th of a unit. At the efficient allocation, a one unit reduction in G combined with a one unit increase in X would leave him equally well off. Consequently, he would increase his well-being by reducing his expenditure on G by one unit of X. Since all consumers find themselves in a similar situation, free market forces would result in a reduction in G below the efficient level, thus creating a market failure.

(a) The efficient provision of local public goods

In a classic article Tiebout (1956) argued that *local* public goods (i.e., public goods for which the perceived benefits are confined to local residents, such as a city park) may be allocated efficiently by consumers 'voting with their feet.' If there is a large variety of communities and consumers are freely mobile, Tiebout's argument predicts that consumers will 'vote with their feet,' and the result will be homogeneous localities in which public goods are efficiently supplied. Since the original Tiebout paper, a considerable body of literature has evolved questioning the Tiebout conclusions.[18] The first problem is that the Tiebout model assumes that the number of separate communities is roughly equivalent to the number of types of consumers, which seems unrealistic. Second, the presence of various sorts of imperfections, especially congestion, is likely to invalidate the Tiebout conclusions. In Frankena and Scheffman (1979), for example, it is shown that some sort of intervention with private markets will generally be required to obtain efficiency.

(b) Non-pure public goods

Before discussing the possible role for provincial (versus local) government in public goods provision it is important to recognize that most goods provided by local and provincial governments are not *pure* public goods. Public roads, for example, have public good characteristics, but since they are subject to congestion in use, they are not *pure* public goods. Average benefits decrease with the number of users because of congestion. Such *impure* public goods create extra difficulties since efficient provision of these goods requires not only a determination of the efficient supply but also policies to influence usage. Finally, some of the common publicly supplied goods may have no significant public good attributes at all, i.e., they may be publicly provided *private* goods. (It has been argued by Stiglitz [1977], for example, that public education is such a good.)

(c) The provincial role in public goods provision

The provincial government has a role in at least two areas of public goods provision: non-local public goods (provincial parks, etc.) and interurban transportation networks and major servicing schemes.

1. *Provincial parks, conservation areas, and wildlife sanctuaries* The economic efficiency arguments for government provision of land for parks, conservation areas, wildlife sanctuaries, and similar uses are rather straightforward compared with those for other areas of provincial land use policy and rest primarily on the public goods and monopoly aspects of such areas.

18 See Flatters et al. (1974), Stiglitz (1977), and Chapter 3.

Consider first the monopoly problem, which may arise in an area which is more or less unique within the relevant commuting range. Then, even if market forces would lead to use of the area as a private park, the allocation of resources might be inefficient. If the owner of the park exercises monopoly power, it could charge an entrance fee above the efficient level and hence restrict use of the park below the efficient level.

Another problem of monopoly power might arise if the ownership of a unique area is initially fragmented. If individual owners hold out for a high price for land, a private entrepreneur might not be able to assemble a park. The government could do so because of its power of eminent domain.

Of course, these monopoly problems would not arise for some types of parks, such as ordinary overnight campground facilities, for which the supply of inputs is elastic and the market is presumably competitive, but monopoly problems could be important for unique areas. Monopoly problems might also be handled by private ownership with government regulation of entrance fees and land assembly rather than government ownership. However, government ownership is also a sensible means of achieving an efficient allocation.

Consider now the public goods argument. Parks, conservation areas, and wildlife sanctuaries have a number of the characteristics of public goods and they provide various positive externalities. For example, provision of land for all of these purposes increases the populations of various forms of wildlife and may prevent some species from becoming extinct. Such effects increase the utility for many people, including people who never visit the area in question, because the wildlife in question migrates and because it appears that some people value wild flora and fauna (particularly the preservation of endangered species) even if they will never see them, perhaps because they are willing to pay to have the option of seeing them. Similarly, the existence of these areas provides flood control, which benefits many people who never visit them.

Because of such public goods and externality considerations, it is likely that too little land would be allocated to parks and similar areas in the absence of government subsidization and/or provision of such facilities. This problem is likely to be exacerbated if the social discount rate exceeds the private discount rate (i.e., if society prefers to have a greater weight given to the consumption of future generations). As an alternative, in some situations the government may achieve the same *allocative* effect simply by prohibiting incompatible uses of the areas in question; however, this alternative involves a very different *distributional* effect when private owners are not compensated.

2. *Interurban transportation networks and major servicing schemes* The importance of the provision of services which facilitate development cannot be overemphasized. Decisions on the location and level of such services often have a more significant impact on the timing, pattern, and volume of development than

policies which attempt to control land use directly. For example, as we shall see in our discussion of the Toronto-Centred Region (TCR) plan in Chapter 7, limited servicing capacity east of Toronto during the recent housing boom helped to contravene the provincial government's avowed 'go-east' policy. Similarly, earlier decisions to locate the Queen Elizabeth Way below the escarpment in the Niagara region and to approve and subsidize servicing schemes in the tender fruitlands conflict with the provincial government's interest in the preservation of these lands. Finally, interurban commuter systems such as GO transit in the Toronto area are likely to encourage 'urban sprawl' and rural development, which again conflicts with stated provincial policies.

The importance of such transportation and servicing schemes is certainly not unknown to provincial officials, but we feel that it justifies much more extensive use of cost-benefit analysis in the decision-making on such public projects. Certainly naïve policy prescriptions such as the Design for Development documents which argued that development should be forced to proceed in a linear pattern so as to minimize transportation and waste removal costs should not be tolerated.

Another important area of provincial involvement in the provision of public goods is subsidization of local public goods. There may be two theoretical grounds for such subsidization. First, there is a body of literature which argues that efficient provision of local public goods and distribution of population between cities may require interurban transfers,[19] which would be best effected by a provincial authority. Second, and of more practical importance, is the fact that some local public goods confer benefits on people who do not reside in the locality in which they are provided. An example is the Toronto subway system, which is used by many non-resident commuters. Local provision of such goods, if financed only by taxes on local residents, is likely to be inefficient.

(d) Agricultural land as a public good

One of the arguments advanced for government protection of agricultural land is based on the contention that agricultural land has some characteristics of a public good, i.e., apart from the benefit of using land in agriculture in terms of output of food, land in agriculture also benefits passers-by and others by virtue of their enjoyment of the open space. This argument will be discussed in Chapter 6.

Uncertainty

Uncertainty is a potential source of market failure for two reasons: (1) there may be insufficient markets or institutional arrangements available for risk-

19 Ibid.

pooling,[20] and (2) market-perceived risks may not be risks from the point of view of the whole economy.[21] In the context of land use policies, the second reason does not seem to be important, so we shall limit our attention to the first.

When insufficient opportunities for risk-pooling exist, agents would be willing to 'sell risks' at a price at which other agents would be willing to buy, but there is no vehicle by which the transactions can be consummated. Such a situation arises from three sources: (a) transactions and other market costs, (b) imperfect information, and (c) moral hazard. Only the first two sources are of importance for our purposes.

(a) Transactions and other market costs

Transactions and other market costs are frictions which impair the (ideal) efficiency of economic exchange. These costs can be important even in the absence of uncertainty, but they are an especially important friction impairing efficient risk-pooling because for many types of risk the number of transactors is too small to bear the costs of operating a formal market. Nonetheless, if transactions, negotiations, and other costs are small enough, non-market transactions arragements are likely to arise, legal contracts being one obvious example.

Because of the inherent heterogeneity of land and housing, land and housing markets tend to be 'thin,' with a resulting loss of efficiency relative to well-coordinated markets such as the stock exchange. Furthermore, the possibilities of risk-pooling are limited relative to many other asset markets. For example, a holder of foreign currency can protect the value of his holdings by trading in the currency futures market, and a holder of major common shares can protect the value of his investment by selling options. However, there is no similar method for a homeowner to insure the value of his equity. (We shall return to this issue later in the chapter when we discuss the causes of government intervention in land and housing markets.)

A case for government intervention on efficiency grounds can be made if such intervention can reduce transactions and other market costs (including the direct and indirect costs of intervention). Perhaps the most fruitful provincial action of this kind would be a simplification of provincial intervention in land and housing markets, particularly in the planning and development process, since current provincial intervention increases transactions and other market costs.

(b) Imperfect information

Informational imperfections impair efficiency because they result in economic agents making 'incorrect' choices. Such imperfections are of particular

20 See Arrow (1963, 1964) and Hirshleifer (1976).

21 See Arrow and Lind (1970).

importance in the context of uncertainty since economic agents may find it difficult to discern accurately the nature of certain risks. For example, insurance companies may have difficulty differentiating between high and low risk customers, which may result in an inefficient menu of policies.

Informational imperfections are certainly present and important in land and housing markets. For this reason considerable resources are devoted to collecting and disseminating information in these markets, particularly by real estate brokers. There has probably also been a significant increase in the quality and quantity of information available in most Canadian land and housing markets in recent years with the advent of the Multiple Listing Service (MLS) system and real estate and trust company surveys.

However, some economists have argued that the private market incentives to collect and disseminate information may result in the amount of resources devoted to such activities being less than the efficient level. The problem here is that the marginal benefits of new information may not be fully appropriable by the agent providing the information because, for example, once a few agents know the information, others are likely to find out about it without having to pay for it. The allocational problem here is similar to the problem of efficiently allocating a pure public good. However, we do not perceive an obvious vehicle for government intervention except for one important case: considerable private resources are devoted to collecting information about current and future actions of the government affecting land and housing markets. Thus there may be a role for the government in increasing efficiency by making information about its actions and plans more widely available.

In particular, a simplification and delimitation of provincial intervention in the planning and development process (as discussed in the Comay, Robarts, and Bossons reports[22]) would reduce the resources now devoted to predicting and modifying the response of provincial agencies to various proposals.

It has been argued elsewhere (Derkowski, 1972, 1975; Markusen and Scheffman, 1977b) that the development control process in Ontario is needlessly complicated, and that it increases the uncertainty faced by economic agents, resulting in higher house prices. We concur in this judgment.

Distortionary government policies

Government policies can introduce distortions which are, themselves, a potential source of market failure. For example, in the absence of other distortions, imposition of a tax on a competitive industry creates a wedge between social and private marginal costs, resulting in a market failure. Obviously there are a myriad

22 Ontario, Planning Act Review Committee (1977); Ontario, Royal Commission on Metropolitan Toronto (1977b); and Bossons (1978).

of provincial government policies which affect land and housing markets. It is important to recognize that the provincial policies which directly regulate land use are often less important in their effects on land and housing markets than are various other policies. We have already argued that decisions on the timing and location of provincially provided public goods (particularly interurban transportation and servicing systems) can have a dramatic effect on the volume and location of development pressures. We shall argue in Chapter 6 that federal and provincial policies such as price supports, taxes and tariffs, and quotas have a significant effect on the rate of conversion of agricultural land.

Finally, many federal and provincial policies affect capital markets, possibly distorting the private discount rate. In particular, it has been argued that corporate taxation and (imperfectly anticipated) inflation increase the private discount rate, which distorts the intertemporal efficiency of resource allocation, possibly speeding the rate of agricultural land conversion, for example.[23] (This issue will be discussed in more detail in Chapter 6.) The existence and effects of existing distortionary government policies must be recognized when benefits and costs of any proposed action are assessed. For example, the effects of agricultural protection must be weighed against capital market imperfections in assessing the efficiency of the existing rate of agricultural land conversion.

Conclusions

We have defined the principal potential sources of market failure (market power, externalities, public goods, uncertainty, and distortionary government policies) and evaluated the case for provincial intervention. Our basic conclusion is negative but important: *the existence of any of the potential sources of market failure is not, in itself, a prima facie case for government intervention.* Thus we must reject the simplistic statements of many planning and other government documents which argue that the obvious presence of externalities, public goods, etc. in land and housing markets is an automatic justification for government intervention. The case for intervention on efficiency grounds must be made on the basis of cost-benefit calculations (including, of course, the direct and indirect costs of intervention).

POLICY INSTRUMENTS

In this section we consider the various policy instruments which could be used for intervention in land markets: (1) direct regulation of land use, (2) tax-subsidy schemes and user charges, (3) government leasing or purchase of limited property rights or title, and (4) changes in liability laws.

23 Bossons (1978).

Direct regulation

Direct regulation is favoured almost to the exclusion of other policy instruments by urban planners. Direct regulation through zoning and planning controls is the traditional method of controlling local location decisions, and (for reasons which are discussed further in the next section) it may be the most efficient method of dealing with local location externalities (although the Houston system raises some doubt). However, the use of planning controls on a broader regional or provincial scale requires further justification as the most suitable instrument. The first defect of direct regulation is that it circumvents the use of markets in determining the allocation of resources (given the existing institutional constraints). Functioning markets generally minimize transactions and negotiations costs. Complete circumvention of the market process may be efficient in some cases, but it seems doubtful that it is efficient in large-scale intervention of the regional or provincial type.

Secondly, direct regulation on a grand scale generally will not achieve its objectives without direct subsidization of location decisions. Planning and zoning controls can restrict land use but cannot ensure that the desired level and pattern of land use will be attained. For example, the TCR plan's avowed 'go-east' policy was unlikely to be realized, even if implemented, without substantial subsidization of location east of Toronto. (The provincial government did provide some small direct encouragement of eastern development by such actions as the relocation of the Ontario Health Insurance Plan offices to Kingston.) The history of the extensive post-war regional planning experience in the United Kingdom reveals both the necessity for and difficulty of subsidization of location decisions to meet the objectives of regional plans.

Finally, because planning and zoning controls create price differentials between land approved and not approved for particular uses (e.g., similar agricultural land approved and not approved for development in growing urban areas), they create economic incentives to circumvent the controls. Such incentives cast doubt on the political stability of controls and (as we shall see in the last section of this chapter) may create inefficiencies.

Tax-subsidy schemes and user charges

Economic theory shows that, in principle, there is always a tax-subsidy scheme which will achieve any particular chosen allocation of resources. This result plus the usual efficiency of markets in minimizing transactions and negotiations costs seemingly gives a theoretical edge to tax-subsidy schemes as a tool for intervention. Indeed, if administrative costs are ignored, appropriate tax-subsidy schemes and user charges set on the basis of marginal social cost are typically the 'first-best' form of government intervention to secure a more efficient allocation of resources in the presence of an imperfection which gives rise to a market failure.

However, the case for the use of these measures for the control of location decisions is weaker on practical grounds. The problem here is that because of the importance of location, land is heterogeneous, resulting in 'the' land market being a market for several (but closely related) goods. For example, the tax-subsidy system equivalent to the zoning controls of any city would be extremely complicated (involving different tax rates on land at different locations) and therefore be likely to incur very substantial administrative costs. Consequently, even though the theoretical literature may treat zoning controls and other direct regulations as 'second-best' policies, this type of policy instrument may actually be the most efficient method of dealing with local locational externalities. Nevertheless, control of the quality of agricultural land converted to urban development could possibly be exercised with a simple system of development taxes, but such a policy is probably not politically feasible.

Therefore the usual economic argument for tax-subsidy schemes as the appropriate tool for intervention is weakened in the case of land markets. However, as our discussion in the preceding section indicated, a suitable tax-subsidy scheme, however informal, will generally have to accompany direct regulation in order to achieve the desired goals in cases of large-scale intervention.

Government leasing or purchase

Government leasing or purchase of land is the obvious and probably efficient method of creating public parks and perhaps greenbelts. Greenbelts can, in principle, be achieved by zoning, but the economic incentives for the owners of greenbelt land to have their land rezoned undermines the political stability of the zoning action. (An excellent example of this problem is the parkway belt around Metro Toronto, discussed in Chapter 7. In contrast, the Ottawa greenbelt has apparently met with more success.)

Because of the powers of eminent domain, it has usually been argued that government is the most efficient agency for assembling large tracts of land for new towns, urban renewal, etc. Numerous examples (including the provincial government's recent experience with the Pickering assemblage) cast some doubt on this assertion. On efficiency grounds, the development industry would seem to be the obvious choice to assemble land. However, political considerations and distributional concerns may make the achievement of efficiency impossible.

Changes in liability laws

As we indicated in our earlier discussion of externalities, liability laws are a potentially important method of efficiently allocating externalities. For small-number externalities, liability laws form the basis of the bargaining position of the parties involved. The government may be forced to intervene in the resulting allocation, even if it is efficient, if the distribution of gains and losses is politically unacceptable. For such situations it may be more efficient to change the liability laws than to intervene directly.[24] For large-number externalities, a carefully designed revision of liability laws to permit certain types of class action suits may be desirable.

THE DEMAND FOR INTERVENTION

Several authors (Stigler, 1971; Posner, 1971, 1974; Peltzman, 1976) have argued that government intervention in the market allocation of resources is generally not motivated by a desire to increase the efficiency of resource allocation. Rather, they argue, such intervention is largely a device to redistribute resources to well-defined constituencies which are then expected to return the favour with votes. Although this theory is overly simplistic in its view of the political process, it is patently clear that most provincial intervention in land and housing markets is not directed to ameliorating market failures. As the analysis developed in this chapter has shown, whether or not a proposed or existing provincial policy can improve efficiency is generally ambiguous, and this ambiguity can only be resolved by an assessment of the costs and benefits of the policy. Furthermore such assessments have not been used as an input in most policy discussions and decisions in the past.

The distributional effects of many policies are, however, clear, a priori.[25] Agricultural price supports and protection clearly redistribute income to farmers and owners of farmland. Homeowners are generally net beneficiaries of policy interventions in land and housing markets. For example, restricting the rate of development through planning and zoning controls, agricultural land conversion controls, etc. makes existing houses more valuable.[26] The present owners of rural non-farm residences and country estates obviously stand to benefit considerably from restrictions on *new* rural non-farm residential development and from development controls in the Niagara Escarpment. Thus, it is not difficult to understand why controls over rural development and agricultural land conversion are popular among the majority of urban voters who own their homes: such controls

24 There is a burgeoning literature on the effects on the structure of liability laws on resource allocation which is beyond the scope of this study. See Polinsky (1978), for example.

25 Although general equilibrium analysis sometimes contradicts the conclusions of partial equilibrium analysis concerning distributional effects.

26 However, we shall show in the next chapter that local government control policies are not necessarily inefficient.

promise to increase the availability of open space, hold down the price of food, and raise the value of existing houses. The political support for land use controls is not built on a foundation of self-sacrifice.

While policies which increase the efficiency of resource allocation at least have the potential to benefit everyone, policies which are primarily redistributive in nature necessarily make some groups worse off, often with an added cost of a reduction in efficiency. Prospective homeowners and renters are the main group adversely affected by development control policies (some farmers and developers are also adversely affected, of course). Given the age profiles of homeowners, renters, and prospective homeowners, much of the government intervention in land and housing markets redistributes income and wealth from future generations to the present generation.

However, such redistribution may not be the only important goal of actual intervention. Earlier we noted that one important difference between land and housing markets and more developed asset markets is the absence of well-coordinated futures markets which would allow agents to insure the value of their holdings. The absence of such insurance markets for a large constituency of risk-averse consumers holding a significant proportion of their wealth in land and housing may be a major source of demand for government intervention in land and housing markets.[27]

Bureaucracy theories of government are also of importance in understanding the roots of the demand for intervention. These theories argue that bureaucrats will promote policies which increase their own well-being (measured by power, prestige, income, etc.).[28] There has been a phenomenal growth in the amount of resources devoted to urban planning in most urban areas of Ontario in the post-war period. This has created a sizable bureaucracy with a vested interest in the use of planning. Much of the local demand for provincial planning and land use controls can probably be traced to this source.

Whatever the source and result of the demand for provincial intervention in land markets, the analysis we have developed in this chapter shows that intervention is not justified on *economic* grounds without a careful assessment of the costs and benefits of intervention and their distributions. Although policy decisions are not made solely on economic grounds, we have argued that such a cost-benefit calculation is an indispensable input into formulation of sensible public policy.

27 A similar argument is made in Breton (1975).

28 Of course, bureaucrats (like other members of society) would normally have developed an ideology to rationalize their activities as contributing to the public good.

3

Provincial land use policies: an introductory survey

We shall now provide an outline of provincial land use policies in Ontario. Almost any policy which has an important impact on the economy will, of course, influence the location of economic activity and land use to some extent; moreover, policies which influence the demand for goods (such as agricultural products or housing) which are produced using land as an input may have a greater effect on land use than explicit land use controls have. However, here we shall be concerned almost exclusively with explicit land use controls. At the end of the chapter we describe briefly a few policies which appear to have important effects on land use even though they do not involve explicit land use controls, leaving a detailed discussion of selected provincial land use policies for the following chapters.

For the benefit of readers who are not familiar with land use planning in Ontario, we have included in Table 3.1 a short glossary of planning terms which are frequently in this study.

OUTSTANDING FEATURES OF PROVINCIAL LAND USE POLICY

Before proceeding with a policy review, it may be helpful to summarize three outstanding features of provincial land use policy:

1. *Increasing intervention* Since the early 1960s there has been a strong trend toward greater provincial intervention in decisions on the allocation of land and land use planning which had previously been left to the private market or local governments.[1] It should be added, however, that recently there have

1 This is true in other provinces as well; see Robinson (1977).

TABLE 3.1

Glossary of planning terms

Term	Definition
Committee of adjustment, land division committee	A committee of adjustment is appointed by a municipal council and has the power to authorize minor variances from land uses permitted by zoning by-laws and the power to grant consents to divide land. A land division committee is the same, except that it operates in a county or metropolitan, regional, or district municipality.
Consent, severance	Under certain circumstances, if a landowner wishes to divide a parcel of land so that part of the original parcel can be sold as a separate lot, he may be able to secure approval (from a committee of adjustment, a land division committee, or the Minister of Housing, depending on where the parcel is located) for what is known as a severance or a land separation consent. Thus, consents and plans of subdivision are alternative methods of approval of division of land.
Minor variance	A change in the use of land which violates the letter but not the intent of a zoning by-law.
Official plan	A document adopted by a municipal or regional council and approved by the Ministry of Housing which states, usually in general terms, development objectives and planning policies for the municipality or region and, typically, designates future uses of land in different areas, sometimes with an indication of the location of future transportation and servicing facilities and the timing of development. Unless amended, land use designations are binding on the municipal or regional government, because the Ontario Municipal Board and courts require conformity of by-laws and public works with the official plan.
Plan of subdivision	If a landowner wishes to divide a parcel of land so that a number of lots can be sold to different parties, he must secure approval from the Ministry of Housing for what is known as a plan of subdivision.

been a few steps in the direction of reduced provincial supervision of local planning, and the province has backed away from comprehensive regional and province-wide planning. If the recommendations of the 1977 Comay and Robarts reports are implemented, there will be a significant further reduction in provincial supervision of local planning, although local governments would be subject to tighter provincial guidelines on protection of agricultural land, rural severances, and development standards and requirements imposed as a condition of subdivision approval.

2. *Absence of explicit policy* With a few significant exceptions, the province generally has not provided statements either of what it conceives provincial interests in land use to be or of its basic land use objectives and policies.[2] Where policies have been announced, they have not always been implemented or prevented the government from taking contradictory actions.

3. *Absence of economic analysis* Within the provincial government, there has been no significant economic analysis of the nature of provincial land use interests or the merits of specific land use decisions or policies. Similarly, with the exception of Bossons (1978), even the most careful studies of provincial land use planning in Ontario do not investigate the economic justifications for provincial intervention in the determination of land use.[3]

We now turn to a review of the major areas of provincial land use planning and control.

PROVINCIAL CONTROL OF LOCAL PLANNING

The Planning Act

The province enacted the Planning Act in 1946 in order both to encourage and to exercise control over local planning. One of the most important characteristics of the Act is the extent to which it retains for the province responsibility for approving and hearing appeals against local government decisions in planning, zoning, and similar matters, and for approving the division of land by plan of subdivision. A number of other acts, including the Municipal Act, the acts establishing the regional municipalities, and the Condominium Act, also give the province responsibilities for approving local land use decisions.

Under the Planning Act, as amended, the province exercises its power through the Plans Administration Division of the Ministry of Housing (formerly the Community Planning Branch of the Department of Municipal Affairs), through the Ontario Municipal Board (OMB), and through the cabinet. The Ministry of Housing is responsible for approval of official plans and plans of subdivision, submits comments to the OMB on zoning bylaws, monitors and has the power to appeal to the OMB decisions of local committees of adjustment and land division

2 The best source of detailed explanations of provincial policies is Estrin et al. (1978).

3 See Adler (1971); Ontario, Ontario Economic Council (1973); Ontario, Planning Act Review Committee (1977); and Ontario, Royal Commission on Metropolitan Toronto (1977b).

committees, and has the power to grant severances and impose zoning orders in areas of the province which are not organized into municipalities, principally in northern Ontario.

The OMB is a virtually autonomous provincially appointed administrative tribunal which is responsible for approving zoning by-laws and amendments, for approving official plans and plans of subdivision upon referral by the Ministry of Housing when they are subject to objections, and for hearing appeals on many types of municipal planning decisions. The OMB has a great deal of power over municipal planning; for example, in the case of appeals it can sustain, reverse, or modify municipal decisions. OMB decisions can be appealed to the cabinet and the Supreme Court of Ontario.[4] In recent years, appeals to the OMB and the cabinet have increased in frequency.

In practice, many provincial agencies apart from the Ministry of Housing and the OMB have come to play a role in the local planning supervision process, particularly in regard to official plan and subdivision approval since the mid-1960s. First, approval by the Ministry of Housing for subdivision plans is typically conditional upon approval by other provincial ministries and by conservation authorities. For example, the Ministry of Transportation and Communications approves proposed new entrances from subdivisions to provincial highways, and the Ministry of the Environment approves proposed sewage systems. Second, other ministries sometimes appeal local planning decisions to the OMB. For example, the Ministry of Natural Resources has appealed severances for cottages which might lead to excessive lake pollution.[5]

One feature of the Planning Act on which all observers agree is that, whatever other effects it has had, the detailed supervision of municipal planning by the province has imposed considerable delays in matters such as zoning amendments and division of land.[6]

The nature of provincial supervision

Provincial supervision of local planning under the Planning Act has not been confined to achieving provincial land use objectives or implementing provincial policies, such as regulating intermunicipal spillovers, protecting agricultural land, restricting urban development in rural areas, or shaping development to conform

4 In 1978 the Supreme Court of Ontario Divisional Court nullified the OMB decision allowing Barrie to annex 9,000 acres from surrounding townships.

5 *Ontario Municipal Board Reports*, 4, pp. 417–18.

6 We argued in Chapter 2 that the complexity of provincial controls over land and housing markets, particularly in the planning and development process, increases transactions and other market costs and impairs the efficient functioning of these markets.

to the Toronto-Centred Region plan. Indeed, in many areas provincial land use policies are either recent or unclear. To some extent, of course, provincial supervision may be motivated by a responsibility to ensure reasonable uniformity of treatment throughout the province in matters such as property rights and the protection of minority interests. However, to a large extent provincial supervision has been designed simply to make sure that local governments and bodies adhere to what are considered to be 'the principles of good planning.' Evidently provincial supervision in the interest of 'good planning' has been motivated by the belief that at least some local governments were not competent and/or lacked the will or political power to plan effectively.[7]

In the following subsections we review ministerial and OMB behaviour in supervising local planning.

(a) Ministerial behaviour

During the life of the Planning Act there have been important changes in the nature and extent of ministerial supervision of local planning to achieve good planning as well as to protect provincial interests.

Until the late 1960s the Community Planning Branch adopted a rather permissive approach to municipal planning. As late as the mid-1960s, Adler found that there was 'no evidence that the Community Planning Branch, to which all (zoning) by-laws are submitted for comment by the (Ontario Municipal) Board before approval, has in the absence of "private" objections ever raised any of its own such as to compel the Board to hold a substantive hearing.'[8]

However, in the late 1960s and early 1970s there was a marked increase in ministerial intervention in municipal planning. Several areas of increasing ministerial intervention will be discussed in Chapters 6 and 7 dealing with rural and agricultural land use controls and regional planning. As another example, until 1973 the effect of planning controls on the supply and price of housing was not considered by the province.[9] However, after the establishment of the Ministry of Housing in 1973 in response to the housing 'crisis' of the early 1970s and the transfer of ministerial supervision to the Plans Administration Division of that ministry, ministerial supervision of municipal planning was also directed toward achievement of the new provincial housing objectives, particularly the objective of increasing the production of housing in the vicinity of Metropolitan Toronto.

7 For further discussion of the motivations of government action, see the last section of Chapter 2.

8 Adler (1971), p. 230.

9 See Ontario, Advisory Task Force on Housing Policy (1973c).

To a large extent, ministerial control over local land use planning is exercised through the power of approval over local official plans. As a condition for approval of official plans, the ministry has required conformity with relevant provincial policies. For example:

1. Most official plans approved since the mid-1960s incorporate severance policies which conform with provincial policy to restrict scattered isolated residential development and ribbon development along highways in rural areas and to confine new permanent residential development in rural areas to existing hamlets and villages, and in special circumstances to estate residential developments.
2. Official plans typically specify that rural residences must be separated from intensive livestock operations such as hog farms by distances laid out in the Ontario Agricultural Code of Practice.
3. Many official plans designate certain areas as 'agricultural' and provide that prime agricultural land is to be preserved for agriculture.
4. Official plans typically require that urban-type development should occur with full municipal services rather than on the basis of septic tanks and wells.

Once such policies are included in official plans, they can be enforced by appeals or comments to the OMB by interested parties, including the ministry.

However, since the drafting and submission of local official plans are voluntary (except for the new regional municipalities), the ministry also relies upon its other powers of approval, comment, and appeal to secure compliance with provincial policies. Appeals by the ministry to the OMB against local planning decisions became common beginning in the mid-1960s, although, as we shall see in Chapter 6, local land use decisions often do not conform to provincial policy in spite of provincial supervision.

Since the early 1970s the provincial government has begun a selective withdrawal from ministerial supervision. The Planning Act was amended to give the Ministry of Housing authority to delegate any of its approval powers to municipal councils, and during the mid-1970s a number of powers were delegated to Metropolitan Toronto and some of the newer regional municipalities, such as authority to comment on municipal zoning by-laws to the OMB, to monitor and appeal to the OMB consents granted by committees of adjustment and land division committees, and to approve plans of subdivision and condominium.

(b) Ontario Municipal Board behaviour

Based on a review we carried out of OMB decisions made during the 1970s, it is possible to make several generalizations about OMB behaviour in the area of local land use control, including situations where the OMB has reversed the decision of a local council or committee.

First, in reaching its decisions, the OMB relies on the statutes if they govern the case. For example, variances or appeals on severances are often decided on the basis that the Planning Act specifies that a zoning amendment or a plan of subdivision is called for in the situation at hand. Most issues which reach the OMB cannot, however, be decided simply by the interpretation and application of laws.

Second, the OMB often relies on conformity with approved or even draft local or regional official plans, zoning by-laws, or land severance policies in deciding on matters such as zoning amendments and severances. However, in many areas of the province no such land use plans or controls exist.

Third, where the matter cannot be decided on the above grounds, as, for example, in the case of approval of an official plan or plan amendment, or in the case of a severance in an area without land use controls, the OMB will attempt to apply any relevant provincial policies. For example, the OMB has decided against official plans which do not conform with the province's Toronto-Centred Region (TCR) plan and has denied severances which conflict with the province's 'Urban Development in Rural Areas' policy to restrict scattered residential development or ribbon development in rural areas.[10] The OMB also decided to allow Barrie to annex 9,000 acres from surrounding townships after the Treasurer of Ontario informed the OMB that this conformed with provincial policy, originally outlined in the TCR plan, for growth of Barrie. (However, this OMB decision was subsequently nullified by the court, which ruled that the evidence of the treasurer was given too much weight.[11]) Conversely, the OMB has allowed rural severances denied by a local land division committee upon appeal by the owner, and has upheld severances granted by a local committee of adjustment despite an appeal by the Department of Municipal Affairs, when the OMB did not believe there was evidence of danger of strip development or scattered residential development which would conflict with provincial policy.[12]

Fourth, where a matter still cannot be decided on any of the above grounds, the OMB falls back upon its judgment concerning the principles of good planning, the proper and orderly development of the municipality, the public interest, or private rights. In doing so, the OMB is often influenced by the opinion evidence of planning consultants. The OMB has denied official plan and zoning amendments which would introduce developments which it felt would not be compatible with existing land uses in the surrounding area, such as a high-rise building in a single-family neighbourhood when it felt the high-rise would

10 *Ontario Municipal Board Reports*, 3, pp. 226–7.

11 *Globe and Mail*, 4 February 1978, p. 1.

12 *Ontario Municipal Board Reports*, 3, pp. 328–9; and 1, pp. 110–12.

adversely change the character of the neighbourhood.[13] Developments have been denied when it appeared they would intensify traffic and pollution problems, or when the adequacy of drainage, water supply, sewers, and parks could not be demonstrated, or when no careful planning study had been carried out.[14] Amendments to official plans which would transfer land from one use designation to another (e.g., industrial to residential) have not been approved when land use projections indicated that a shortage (i.e., an excess of projected 'requirements' over availability) was more likely to appear for the original category of land than for the new one. Conversely, in appeals the OMB has compelled municipalities to accept changes when the OMB did not believe they conflicted with any of the principles of good planning.[15] Similarly, the OMB has rejected objections from provincial ministries to local land use decisions when the OMB felt that 'all reasonable planning grounds, the planning opinion-evidence and the consideration of the public interest' were on the side of the municipality.[16]

It is, of course, in this fourth situation that the OMB exercises the greatest degree of discretion in its decisions and essentially makes provincial land use policy by default. Thus, Cullingworth (1978, p. 33) remarks that the OMB 'now has the role of deciding upon many of the more difficult political planning issues which elected representatives are glad to shirk.' What must be remarked upon here is that there are many such areas of land use control in which the provincial government has not in fact given policy guidelines to the OMB and where the OMB has, as a result, evolved its own policies over the years.[17]

Finally, it should be added that the OMB is passive in its approach to protection of provincial planning interests. For example, according to Adler, the OMB gives automatic approval to zoning by-laws unless they are opposed by an interested party, rather than considering on its own initiative whether the proposed by-laws are in conformity with provincial interests.[18]

13 *Ontario Municipal Board Reports*, 1, pp. 162–3. According to Estrin et al. (1978), pp. 369–71, under the chairmanship of J.A. Kennedy from 1960 to 1972, the OMB often decided against local governments which approved developments that would interfere with the amenities of established neighbourhoods. However, after 1972 the OMB changed direction and it became more difficult for citizens' groups to prevent developments approved by municipal councils.

14 *Ontario Municipal Board Reports*, 7, pp. 91–5.

15 *Ontario Municipal Board Reports*, 3, pp. 117–22.

16 *Ontario Municipal Board Reports*, 5, pp. 117–25.

17 The OMB is not bound by precedent, and its decisions depend to some extent on which members of the OMB are involved.

18 Adler (1971), p. 230.

(c) Reports of the Comay Committee and Robarts Commission

Suggestions to reduce provincial supervision of local planning have been made on a number of occasions,[19] and in 1977 two major provincially sponsored studies recommended a substantial reduction in provincial control over local planning.

The Planning Act Review (Comay) Committee recommended greater municipal autonomy in planning, with provincial involvement limited in scope to specific provincial interests (conservation and management of the natural environment and rural land, provision of housing, etc.) and in means (formal vetoes, etc.). Municipal councils would be assigned the final authority over all their own planning instruments (official plans, zoning by-laws, subdivision plans, land separation consents, etc.), subject to appeal procedures and provincial vetoes for stated reasons. Existing requirements for provincial approval would be ended.[20]

The Royal (Robarts) Commission on Metropolitan Toronto recommended a reduction of provincial supervision of municipal planning: 'The planning process should maximize local autonomy by limiting the review and approval powers of bodies more remote from the local political scene than the local council. While there is no question that review powers are necessary to ensure that local planning meets the needs of the broader society and does not violate individual rights, it is no longer necessary for provincial authorities to review and formally approve municipal planning actions which are acceptable locally and are properly local in nature.'[21]

The rationale for the Comay and Robarts recommendations to increase municipal autonomy in planning was largely to increase political accountability in municipal planning. Bossons (1978) argued that these two reports did not give adequate weight to the importance of 'quasi-constitutional' restrictions on municipal decision-making (e.g., the requirement of conformity of municipal by-laws to local official plans, the right to appeal municipal decisions to the OMB) in reducing uncertainty and protecting individual rights and minority interests.

It is important to point out, however, that in two areas the Planning Act Review Committee recommended new restrictions on municipal autonomy in planning in order to protect provincial interests. First, the committee recommended that the province impose a base-level rural zoning by-law and consent policy to control development in rural areas. Second, the committee recommended that the province control development standards and requirements

19 See Ontario, Ontario Economic Council (1973).

20 Ontario, Planning Act Review Committee (1977).

21 Ontario, Royal Commission on Metropolitan Toronto (1977b), vol. 2, p. 215.

imposed by municipalities as a condition for approval of new subdivisions in order to limit the power of municipalities to 'sell' development approvals for the financial benefit of present residents and to exclude low income and other types of residents.[22]

PROVINCIAL REGIONAL PLANNING

The 1960s saw not only an increase in provincial supervision of municipal planning but a new policy of direct provincial responsibility for comprehensive regional planning. Provincial regional planning began with the Metropolitan Toronto and Region Transportation Study (MTARTS) and was institutionalized in 1966 in the Design for Development program, which was placed under the responsibility of the Regional Development Branch of the Ministry of Treasury and Intergovernmental Affairs (later the Ministry of Treasury, Economics and Intergovernmental Affairs [TEIGA]).

The first major product of the Design for Development program was a 1970 plan for the Toronto-centred region,[23] which was based on one of the alternative development patterns suggested by the MTARTS report.[24] As we shall see in our analysis of the TCR plan and Ontario's experience with provincial regional planning in Chapter 7, although the plan was declared to be provincial policy, the government did not follow through on many of its features, and the province has now abandoned much of the Design for Development approach to comprehensive regional planning.

DIRECT PROVINCIAL PLANNING

The Planning Act not only gives the province powers to control planning decisions by municipal bodies but also gives the Ministry of Housing the power to control land use directly by imposing zoning orders in any area of Ontario, subject to conformity with local official plans and subject to hearings before the OMB. Also, as we shall see below (p. 47), the Public Lands Act gives the Ministry of Natural Resources the power to control land use directly in areas of northern Ontario outside municipalities through a system of development permits.

Recent legislation gives the province much more extensive powers to carry out planning and impose land use controls directly and in the process to bring

22 Ontario, Planning Act Review Committee (1977), pp. 33–4, 128.
23 Ontario, Government of Ontario (1970b).
24 Ontario, Department of Municipal Affairs, Metropolitan Toronto and Region Transportation Study (1968).

about the amendment of municipal planning and zoning decisions which have already been approved if they do not conform to the provincial plan. In 1973 the provincial government enacted the Ontario Planning and Development Act, which gave broad planning and development control powers to the province.[25] The province assumed the power to declare any area of Ontario a 'development planning area' and to prepare a development plan for the area, subject, however, to detailed provisions for consultation with local interests, public hearings, and subsequent review. It also assumed the power to expropriate any land in the area for the purposes of implementing such a development plan. Local governments were prohibited from enacting by-laws or taking actions in conflict with such a development plan, and the provincial plan would prevail over local official plans and zoning by-laws if there was a conflict.

As of 1979 the Ontario Planning and Development Act had been used only in the Parkway Belt West near Metro Toronto. However, in 1973 the province also passed the Niagara Escarpment Planning and Development Act, which initiated a second special purpose provincial land use planning exercise. Apart from the distinction that the Ontario Planning and Development Act was applied to the Parkway Belt West but not to the Niagara Escarpment, distinguishing features of the Niagara Escarpment planning exercise were that the plan was to be drafted by a provincially appointed Niagara Escarpment Commission rather than TEIGA, and that implementation of the plan was to be achieved by development control rather than zoning. The activities of the province in planning for the Parkway Belt West and the Niagara Escarpment are discussed in Chapter 7, which deals with regional planning and the Toronto-Centred Region plan.

PROVINCIAL NEW TOWNS

The province has also assumed responsibility for the planning and development of new towns in situations where a large increase in employment is expected to lead to rapid urban development outside of existing urban areas. The planning has occurred in two phases. First, during the 1950s and 1960s the province took the initiative in planning and developing several new mining towns in northern Ontario. In 1954 the Mining Act was amended to provide that the surface rights of mining claims could be reserved by the province for development of townsites. The provincial Community Planning Branch, then in the Department of Planning and Development, was given responsibility for planning the new towns. The planning of the first of these new mining towns, Manitouwadge, was begun

25 In the Planning Act, the responsible ministry is Housing. In the Ontario Planning and Development Act, the responsible ministry is TEIGA.

in 1954.[26] This was followed by new towns at Elliot Lake, Ear Falls, and Temagami. Since 1970, however, provincial policy in northern Ontario has encouraged growth of existing towns rather than development of new ones.

Second, during the early 1970s, the province undertook not only planning but large-scale land acquisition and responsibility for development for four large new cities in southern Ontario. By 1977 the Ontario Land Corporation had acquired approximately 57,000 acres of land for these four new towns: South Cayuga (12,690 acres at a cost of $31.5 million) and Townsend (13,500 acres, $36.5 million) in the Regional Municipality of Haldimand-Norfolk, North Pickering (20,084 acres, $235.5 million) northeast of Toronto, and Edwardsburgh (10,425 acres, $8.8 million) adjacent to the town of Prescott in eastern Ontario.[27] In the case of North Pickering, in 1972 the province announced plans to develop a provincially sponsored new town for 150,000–200,000 (later reduced to 70,000–90,000) people. However, despite large-scale provincial land acquisition for all four cities and elaborate provincial planning exercises for Townsend and North Pickering, as of 1979 there was considerable doubt whether and when actual development would take place. Moreover, by 1977/78 the Ontario Land Corporation had adopted a 'policy of minimizing land purchases.'[28]

LAND USE ACTIVITIES OF THE MINISTRY OF NATURAL RESOURCES

The Ministry of Natural Resources has responsibilities for land use in extensive portions of the province. Under the Provincial Parks Act, the ministry manages 122 provincial parks on 10,772,000 acres and 105 provincial reserve areas on 1,257,000 acres. Under a recent amendment to the Act (1976), the ministry is currently drawing up a statement of policies to direct future development of the provincial park system.

Under the Conservation Authorities Act, the ministry funds and exercises approval powers over conservation authorities, which are composed of representatives of municipalities. Each of the 38 authorities has jurisdiction over a single watershed or a number of smaller ones. The authorities restrict urban development on and otherwise regulate the use of floodplains and river valley

26 Ontario, Department of Planning and Development (1957).

27 Canada, Environment Canada, Lands Directorate (1977), pp. 74–6, and Ontario, Ontario Land Corporation (1977/78), p. 11. Cost figures include carrying costs through 31 March 1977.

28 Ontario, Ontario Land Corporation (1977/78). p. 7.

lands, and they have acquired over 100,000 acres of such land in nearly 300 conservation areas and developed it for recreational use.

There are about 200 million acres of provincially owned crown lands in Ontario. The permissible uses of crown lands are sufficiently broad so that the provincial government may license, lease, sell, or otherwise allocate surface, mineral, and/or timber rights to municipal governments (e.g., as easements for utilities or transportation facilities), to individuals (e.g., as housing sites, farms, or cottage lots), and to private companies (e.g., for prospecting, mining, logging, or tourist developments).

A few areas in northern Ontario are organized into municipalities or are subject to zoning orders imposed by the Ministry of Housing. However, the greatest part of northern Ontario is unorganized and unzoned and falls under the Ministry of Natural Resources, which has the power to exercise land use controls under the Public Lands Act through a system of development permits, which can be applied to all land uses (except mining) which involve buildings or improvements.

Under the Pits and Quarries Control Act (1971), the Minister of Natural Resources controls the operation of existing and proposed pits and quarries through the issuance of licences, subject to provision for hearings before the Ontario Municipal Board. Licences must be renewed annually and may be refused for environmental reasons, in order to ensure that a glut of the material will not occur, or to hold down traffic on roads. The Act also levies a small security deposit for post-extraction rehabilitation of pits and quarries.

There are 132,000,000 acres of crown land forests. In addition, under the Forestry Act the province manages 263,000 acres of land owned by other levels of government, and under the Woodlands Improvement Act the province has agreements with private landowners covering 242,000 acres. Under the Game and Fish Act, there are 41 wildlife areas throughout the province on 80,000 acres of provincially owned land and on 21,000 acres of land owned by private parties and other public agencies.

The Ministry of Natural Resources also exercises controls over land use under the Wilderness Areas Act and the North Georgian Bay Recreational Reserve Act, and the Ministry of Culture and Recreation is responsible for parks created under the Historical Parks Act.

LAND USE ACTIVITIES OF THE MINISTRY OF THE ENVIRONMENT

Regulations designed to protect the environment often have the effect of land use controls. Under the Environmental Protection Act, the Ministry of the

Environment regulates private sewage disposal systems (e.g., cesspools and septic tanks), solid waste disposal sites, and air emissions. Under the Ontario Water Resources Act, the ministry has responsibilities for water supply and sewage disposal schemes.

The combination of the power to restrict residential development on septic tanks and the responsibility for major servicing schemes gives the province dramatic powers to control the timing and location of urban development. Thus, given the provincial policy of restricting urban development to fully serviced subdivisions, the timing and location of servicing schemes are important determinants of the pattern of urban development in the province. For example, in the Toronto-centred region, development west of Metropolitan Toronto was possible in the early 1970s because of the Peel servicing schemes of the 1960s. By contrast, development north and east of Toronto was awaiting the York-Durham servicing scheme, except in areas such as Markham which were connected to the Metro Toronto sewer system.

The Environmental Assessment Act, which began to come into effect in 1977, requires the submission to the Ministry of the Environment of an assessment of the environmental impact of governmental and 'major' private undertakings. Under the Act, approval of a prospective undertaking may be granted directly by the minister if there is no public objection, or after a review by the Environmental Assessment Board if there is.

Potentially, the province could exercise considerable control over land use under the authority of this Act. In practice, many undertakings have been exempted from the Act. According to one report,

> The list [of exemptions] which takes up 63 pages of fine print in the *Ontario Gazette*, exempts everything from highway developments to expansion of any college or university. It exempts more than 150 sewage and water treatment plants, almost all government property leasing and housing development; it exempts industrial parks, provincial park master plans, mining exploration, all Natural Resources Department dams and dikes, all Toronto Area Transit Operating Authority (GO-train) developments, endless miles of Ontario Hydro power lines and power plans, and even the government's Tourism Development Plan. Officially the reason so many projects were exempt is that they were already on the drawing board. But some of the projects are years away. Some are not even planned, beyond a name and a vague location.[29]

Apart from these exemptions, housing is exempt from the Act, and municipal governments have been exempted until a study of the implications of the Act,

29 *Toronto Star*, 29 April 1978.

including its relation to the Planning Act, is completed. As of 1977, the Act had been applied to only two private projects.[30]

LAND USE ACTIVITIES OF OTHER PROVINCIAL AGENCIES

As we indicated in Chapter 2, many government policies other than explicit land use controls have important effects on land use in the province. For example, the federal and provincial governments have a number of policies which influence demands and production costs for agricultural goods and housing. Such policies change the derived demand for land for agricultural and residential use, and hence affect land prices and the rate of conversion of land from agricultural to other uses. Important policies of this type include federal import restrictions on agricultural products; federal-provincial agricultural price support policies; federal-provincial subsidization of crop insurance; provincial legalization of agricultural marketing boards; provincial policies of assessing farms at a lower rate than non-farm properties, rebating property taxes on farms, and exempting family farms from the succession duty; and federal-provincial housing subsidies.

A second set of policies which have important effects on land use are provincial policies for the provision and subsidization of major public services which are generally complements of residential land use, particularly sewage systems and commuter transportation facilities.[31] The role of these facilities in affecting the timing, location, and density of residential development was stressed in Chapter 2.

Thus, apart from specific land use control and planning activities, a number of provincial ministries and agencies are responsible for programs which may have important impacts on land use, such as the Ministry of Housing (Ontario Housing Corporation land assembly and development[32] and other housing programs), the Ministry of Transportation and Communications (highways, GO transit, transportation subsidies), the Ministry of Agriculture and Food (marketing boards, the federal-provincial ARDA program of farm enlargement and consolidation), the Ministry of Industry and Tourism (Ontario Business Incentives Program for northern and eastern Ontario), Ontario Hydro (transmission lines),

30 The proposed INCO power dam on the Spanish River and the proposed Reed Paper mill at Ear Falls (Estrin et al., 1978, p. 45).

31 There are also federal subsidies for some of these projects, e.g., CMHC grants for waste control projects and a $9.7 million federal grant to GO transit in 1975.

32 As of 1977, the Ministry of Housing and the Ontario Housing Corporation owned approximately 23,000 acres of land intended for eventual development for housing, apart from land in the new town sites discussed above. See Canada, Environment Canada, Lands Directorate (1977), pp. 71–2.

the Ontario Energy Board (pipelines), the Liquor Control Board (protection of domestic wines), and the Ministry of Revenue (land transfer tax, property assessment, farm tax rebate).

SUMMARY OF POLICY TRENDS

We have seen that provincial intervention in land use planning and the allocation of land increased greatly after the early 1960s. The decade from the mid-1960s to the mid-1970s saw announcement of an 'Urban Development in Rural Areas' policy, the Design for Development program, and the Toronto-Centred Region plan, the Parkway Belt and the Niagara Escarpment planning programs, land assembly for four major provincial new towns, and a policy of preserving agricultural land. In addition to such dramatic developments, there was an increase in provincial supervision of local planning and an increase in the frequency with which land use issues were appealed to the OMB.

By the late 1970s, the trend appears to have changed. The province has scaled down its objectives in a number of areas, most notably comprehensive regional planning, it has shown a willingness to compromise on matters such as the extent of the parkway belt and escarpment planning areas, and it has generally refrained from taking new initiatives. There is also an incipient trend toward reduced provincial supervision of local planning, apart from the provision of guidelines in areas of provincial interest such as rural severances and agricultural land.

4

The provincial role in municipal planning

The provincial interest in municipal planning and the local subdivision approval process arises from three legitimate sources: (1) responsibility for non-local issues such as non-local externalities and preservation of agricultural land; (2) responsibility for protection of individual rights and minority interests; and (3) the possibility that the municipal planning process is itself a potential source of market failure, i.e., that municipal development controls confer potential market power on municipalities. These legitimate sources of provincial concern may serve as a partial justification for the greatly increased direct involvement by the province in the monitoring and control of municipal planning and the local subdivision approval process in the post-war period.

In recent years provincial involvement in local affairs has been subjected to a substantial volume of criticism. Some of the criticism apparently arises mainly from the evident increasing sentiment for decentralization of political authority in Canada, and in its extreme form such criticism argues for complete local autonomy. However, the three legitimate sources of provincial interest in the municipal planning and local subdivision approval process enumerated above militate against complete local autonomy on efficiency and equity grounds. In the following sections we shall discuss some of the major specific criticisms of provincial involvement in its present form and evaluate some of the recently proposed changes in the distribution of powers and mechanism of provincial involvement.

THE COMPLEXITY OF THE PLANNING AND SUBDIVISION APPROVAL PROCESS

In its present form, provincial supervision is needlessly complicated and detailed and excessively concerned with making sure that local governments and bodies adhere to what are considered to be 'the principles of good planning' even on matters of purely local interest. Several authors have commented on the extreme complexity of the subdivision approval process in Ontario (e.g., Derkowski, 1972, 1975; Markusen and Scheffman, 1977b), and this problem has been a major concern in three reports by Comay and in reports by the Robarts Commission and Bossons.[1] This complexity has been widely cited as contributing to higher housing prices and obscuring political accountability for planning decisions. It is apparently agreed in virtually all quarters that the level of complexity should be reduced by giving municipalities greater autonomy in local planning and limiting provincial constraints to the protection of provincial interests. We concur that such a development would be desirable. In certain types of cases, to prevent inordinate delays it might be desirable to set maximum limits on the time the government is allowed to make a decision following an application by a developer. These measures would, among other things, promote a more efficient functioning of land and housing markets and a more efficient allocation of resources.

PROVINCIAL GUIDELINES

In the past the provincial government has failed to provide coherent statements of its land use policies. It has thus been very difficult for local authorities to enact policies which satisfy provincial concerns. The few statements and guidelines that have been issued have been too vague to provide developers, municipal planners, or the OMB with much guidance on specific issues. Recently, the government has attempted to remedy the problem to some extent through statements of policy such as the *Green Paper on Planning for Agriculture*. However, usually these statements have simply enumerated provincial interests in land use without providing either specific directions for dealing with particular cases or rudimentary trade-off valuations with which municipal authorities could measure conformity of possible policies with provincial objectives (such as the maximum cost the province thinks should be borne in order to 'save' an acre of prime or unique agricultural land). The existing enumerations of provincial

1 Ontario, Ontario Economic Council (1973); Ontario, Advisory Task Force on Housing Policy (1973c); Ontario, Planning Act Review Committee (1977); Ontario, Royal Commission on Metropolitan Toronto (1977b); Bossons (1978). See also Canada, Federal/Provincial Task Force on the Supply and Price of Serviced Residential Land (1978), especially pp. 38–43, which makes an important distinction between the length and the restrictiveness of the approval process.

interests typically include such things as preservation of agricultural land, conservation of natural resources, protection of the environment, provision of housing, reduction of regional disparities, provision of interurban transportation and servicing schemes, and regulation of intermunicipal spillovers and intermunicipal disputes.

In this important matter of understanding what provincial interests are, we feel that economic analysis should play an explicit and significant role. As stated in Chapter 2, although economic arguments are not the only determinant of policy decisions, economic analysis is indispensable to rational policy-making whenever the allocation of scarce resources or the distribution of income and wealth are affected, as in the case of land use policies.

Thus, any attempt to enumerate provincial interests in land use should be based, among other things, on the economic methodology explained in Chapter 2 for identifying *possible* justifications for government intervention in resource allocation based on potential sources of market failure. However, as we emphasized before, the existence of any of the potential sources of market failure is not, in itself, a prima facie case for government intervention. Rather, the case for provincial intervention on efficiency grounds must be made on the basis of cost-benefit calculations.

Cost-benefit calculations should play an explicit and significant role at two stages in land use policy formulation. First, before adopting any policy for general application, the province should consider its costs and benefits and their distribution among the population, and the results of this evaluation of costs and benefits should be presented to the public as part of the justification for the policy. Second, any policy intended for general application may be inappropriate in some particular circumstances. Consequently, there should be explicit provision for appeals against general policies in specific circumstances based on various grounds. One of the bases which should be recognized for appeal against general provincial policies in specific circumstances should be that the costs exceed the benefits of the policy in the case at hand; that is, variances could be allowed where they would increase the efficiency of resource allocation.

In the context of appeals to the province against municipal planning decisions, we also recommend that economic analysis be given a more explicit and significant role. Correspondingly, a lesser role should be given to the simplistic 'principles of good planning' promoted by urban planners in general and by planning consultants presenting evidence at the OMB in particular. Thus, consideration of benefits and costs should be a routine part of OMB hearings and an important basis for OMB decisions, and provincial decisions to veto, reverse, or modify municipal decisions should be accompanied by an explicit statement of the economic benefits and costs.

PROVINCIAL CONTROL OF THE LOCAL DEVELOPMENT CONTROL PROCESS

Our discussion will be concerned with the equity and efficiency issues which may arise at the provincial level from the exercise of development control powers by municipalities. We shall focus on two basic issues that arise in this context: (1) is it fair or efficient for municipalities to treat new-comers differently from existing residents? and (2) what are the efficiency implications of the control of development by municipalities? To address either of these issues requires a framework within which local development control policies are explained, i.e., what are the major incentives for municipal control of development and what sort of development controls will result? Put another way, the question of whether the development control policies of municipalities are equitable or efficient requires an explanation of why and in what form such policies may arise. Although the possible effects of development control policies (on equity and the price of housing, for example) have been widely discussed, virtually all discussions have lacked a coherent framework which can explain why and how municipalities will control development. Elsewhere[2] we have attempted to provide such a framework, and in what follows we describe the implications of this analysis.

The efficiency of local development control policies

When employing the economic methodology described in Chapter 2, the first question to be addressed is what the potential sources of market failure arising in this context are. We shall first consider whether municipalities have potential market power because of their development controls.

(a) Do municipalities have potential market power?

In our analysis we distinguish between what we define as 'small' and 'large' cities. A small city is one in which the effects of development control policies on non-inhabitants of the city are insignificant. A large city is one in which such effects are significant. For example, if Woodstock, Ontario, were to prohibit all future development, the effect on the rest of Ontario would presumably be insignificant. However, if the metropolitan Toronto area had adopted such a policy in the post-war period, it is unlikely that the effects on the rest of Ontario would have been negligible.

It is important to be clear about what is meant by large and small in this context. Development control policies affect the net number of *new* inhabitants

2 Frankena and Scheffman (1979).

in a city. Thus, although Metropolitan Toronto was not a large city in terms of total population at the beginning of the post-war period, the net number of new inhabitants during the post-war period was a significant proportion of the total Ontario population. In our terminology, then, Metropolitan Toronto probably must be considered a 'large' city. In contrast, Montreal is a large city in terms of total population, but during the last 10 years the net number of new inhabitants was fairly small (and would have been even in the absence of significant development controls). Thus in our terminology Montreal would probably have to be considered a 'small' city.

This distinction between large and small cities is important because small cities do not have potential market power. In order to have potential market power the city's actions must have a non-negligible effect on non-inhabitants. In popular discussions of urban housing markets it is generally assumed that all cities (and even neighbourhoods) have potential market power. The evidence on which this assumption is commonly based is the differential price of housing and urban land in different locations. For example, the fact that seemingly similar houses in different subdivisions or cities have different prices is often cited as evidence of the exercise of market power (although these arguments are not generally couched in these terms). However, these arguments usually misperceive the cause of the price differential. Locational advantages, amenities, local job market opportunities, etc. are critical factors determining the price of houses. The fact that similar houses at different locations have different prices is generally due to these factors, not to the exercise of market power by either the development industry or the municipality.

Most cities in Ontario would have to be considered small by our definition, Metropolitan Toronto being a likely exception. Other possible exceptions include some of the new regional governments created in the past decade. For example, the amount of new development which has occurred in the Regional Municipality of Peel in the last 10 years could qualify Peel as a large entity. In our opinion the exercise of market power by *municipalities* is not an important problem in Ontario. Most municipalities are too small to have significant market power, and the planning authorities of the potentially large entities are too fractionated to achieve the coordination of policy necessary to exercise market power.

In our opinion a source of greater market power is *provincial* involvement in local development controls. For example, since homeowners are the largest political constituency actively concerned with development controls, it could perhaps be argued that this constituency has influenced the provincial government to exercise market power in the control of development through its control of the provision of major servicing schemes and transportation networks. Conspiracy theorists might prefer this explanation for the shortfall in servicing capacity in the Toronto area and some other parts of the province during the recent housing boom. We would discount this argument, but the potential for the provincial government to exercise market power combined with the interest of a large constituency (homeowners) reinforces our discussion in Chapter 2, which argued that provincial decisions on major servicing schemes and transportation networks should be placed in a more sensible framework in which the costs and benefits and equity implications are fully assessed.

Although we have argued that the exercise of market power by municipalities is not an important problem in Ontario, local development control policies may still justify a provincial interest on grounds of economic efficiency, for the reason that municipal authorities may not have the proper incentives to achieve an efficient allocation of resources. To evaluate this possibility we must first determine what the economic incentives for municipal development control policies are.

(b) What are the economic incentives determining the development control policies of municipalities?

Although there are a myriad of concerns (of both the electorate and the government) which affect the outcome of any political process, we shall restrict ourselves here to the major economic incentives for municipalities to control development. Such restriction is appropriate, we believe, because we shall be concerned only with the *aggregate* properties of development controls, i.e., with how much and with what type of new development should be allowed. We shall not consider problems which are more local (to a particular neighbourhood), such as whether high-rise development should be allowed adjacent to low-density residential development; these issues have been adequately discussed elsewhere, in the extensive literature on the theory of zoning.

There are three major segments of the electorate in a city which evidence an important concern (through attendance at planning board and city council meetings) with the *aggregate* properties of development controls: homeowners, developers, and the business community. In our experience, renters, as a group, evidence much less active concern with this issue, perhaps because they have a smaller stake in the community.

1. *Homeowners* Homeowners are concerned with the aggregate properties of new development because it may affect their well-being through its effect on their incomes, house values, taxes, the services provided by the city, and the aggregate external effects of new development, such as increased congestion. Average 'net' income (net of congestion costs, etc.) of homeowners is likely to be reduced by new development, *ceteris paribus*, because of the depressing effect

of increased labour supply and increased travel congestion costs, etc.[3] To the extent that net income opportunities are capitalized in house values, homeowners' house values will be affected in a manner similar to their net incomes.[4] Finally, the property tax system obviously creates an interest by ratepayers in new development, because of its effect on the tax base. New development is thus likely to impose a cost on homeowners through its effect on net income and house values, but new development may confer a benefit on homeowners through the broadening of the tax base (especially if, as has been true in the past, new-comers can be taxed at a higher effective rate).

2. *Developers* Developers, of course, favour a minimum of restriction of new development, except to the extent that such restriction can be directed against the existing potential competition. Because most Ontario cities are 'small' in the sense of our earlier definition, and because the empirical evidence indicates that the development industry is not highly concentrated in Ontario cities, the chance that the development industry in a city could exercise market power through influencing the planning authority or city government is, in our opinion, remote.

3. *The business community* The increased markets and labour supply resulting from new development are strong incentives for the business community to be pro-development. This incentive is somewhat tempered to the extent that businesses must bear a larger tax burden because of new development.

(c) What sort of development control policy will arise from these economic incentives?

Elsewhere[5] we have developed an economic theory of development controls that exhibits the sort of development control policies which will arise from the economic incentives described above. We shall now summarize the results of that theory. To the extent that development control policies are determined in a democratic manner, since homeowners are by far the largest constituency actively interested in the aggregate properties of development controls, their preferences will dominate. The economic costs and benefits of new development accruing to homeowners will generally result in a net incentive to control development. Furthermore, the property tax system itself provides a strong incentive on the part of all ratepayers to support a policy of house- and lot-size zoning, in order to maximize the tax base for a given population. Therefore the existence

3 See Chapter 2 and Richardson (1973).
4 Frankena and Scheffman (1979).
5 Frankena and Scheffman (1979).

of the economic incentives we have described leads us to predict use of the sort of aggregate development control policies that we actually observe in most urban areas: control of the level of development and the use of lot- and house-size zoning.

(d) What are the economic efficiency implications of these development control policies?

Development pressures are a potential source of market failure to the extent that new development creates externalities affecting the existing residents (e.g., increased congestion in travel and the use of local public goods). In the absence of any controls, the economic incentives for a newcomer to locate in a given urban area depend on the perceived benefits and costs accruing to him. He, of course, will not directly bear the external costs (in the form of increased travel time, more congested schools, etc.) which he will impose on other residents. Consequently, in the absence of controls the economic calculations of new-comers concerning the net private benefits of a given locality will result in the pattern and level of new development being economically inefficient. Therefore there is a possible economic justification for municipalities to control development.

However, we have shown elsewhere[6] that to the extent that homeowners' preferences dominate the determination of development control policies, such policies are unlikely to result in an efficient allocation of resources. This is because the property tax system creates incentives for inefficiency, since the incentive under a property tax system is to maximize the tax base, for any given population target, which is not efficient.[7]

In summary, there is a possible economic justification for municipal development control policies on efficient grounds, but to the extent that homeowners' preferences dominate the determination of policies, the policies that are chosen by the municipality are unlikely to attain full efficiency. The question therefore arises as to whether it is appropriate on these grounds for provincial authorities to control the local development control process, a question to which we shall now direct our attention.

6 Frankena and Scheffman (1979).

7 The efficiency of development control policies under a head tax and property tax system is analysed in Frankena and Scheffman (1979). The externalities created by new-comers are a potential source of market failure, and the market failure which results occurs in the land market. One problem with the property tax system is that it creates incentives to intervene in the *house* market (through house-size zoning), which is not an efficient policy.

THE RECOMMENDATIONS OF THE PLANNING ACT REVIEW COMMITTEE – A CRITICAL REVIEW

The definition of the provincial role in municipal planning

Most commentaries on the current state of the planning process in Ontario, including the Report of the Planning Act Review Committee (PARC), have urged a reduction of the provincial role and increased local autonomy. The PARC Report recommends that

> the provincial interest in municipal planning should be formally defined to comprise:
>
> Implementation of provincial policies and programs in economic, social and physical developments, protection of the natural environment and management of natural resources; and the equitable distribution of social and economic resources.
>
> Maintenance of the provincial financial well-being.
>
> Ensuring civil rights and natural justice in the administration of municipal planning.
>
> Ensuring coordination of planning activities of municipalities and other public bodies, and resolving intermunicipal planning conflicts.[8]

Thus in most cases PARC recommends that the province confine itself to issues which have a stated provincial interest (e.g., preservation of prime agricultural land), or problems in which inequities may arise.

We strongly concur with the view that provincial responsibilities in the planning process should be clearly defined, simplified, and reduced from their present level. However, we would stress the importance of the efficiency of resource allocation as a legitimate provincial concern. The formulation of provincial guidelines and their implementation should, in our opinion, be made in a framework in which the economic impact of proposed policies is fully considered. We would suggest a formal system of requiring 'economic impact statements,' both as a fundamental instrument in the process of the formulation and change of provincial policies and as evidence necessary for an adequate review of proposed local deviations from provincial guidelines. These economic impact

8 Ontario, Planning Act Review Committee (1977), p. i.

statements would assess the effect of a proposed (or existing) action on equity and efficiency (using the framework suggested in Chapter 2). We believe that the PARC recommendations neglect the importance of the efficiency of resource allocation as an issue of provincial concern, overemphasizing the (legitimate) provincial concern with equity. In policy formulation both equity and efficiency implications of proposed actions must be considered.

Finally, we cannot concur with some of the PARC Report's specific recommendations on the setting of provincial development standards and requirements, an issue which we shall consider next.

The provincial role in setting development standards and requirements

Of the seven major recommendations by PARC in the area of development standards and requirements, we disagree with two in their present form:

The Minister should be required to establish, through regulations, the range of development standards and requirements that can be incorporated in municipal planning instruments and imposed on development and subdivision proposals. On the basis of established health, safety and amenity needs, he should be required to set an upper limit on minimum lots and floor area and a lower limit on maximum density to be established in major areas of new development, and the maximum servicing requirements and engineering standards that can be imposed in such areas. The Minister should be authorized to alter these standards and specifications at the request of a municipality, for stated reasons.

Municipalities should be allowed to impose only those financial levies which relate 'fairly and reasonably' to the particular development involved. Levies should be purpose-specific only, rather than general. Developers and subdividers should not be required to pay for the capital cost of physical works beyond what is required to bring the site or subdivision to a condition suitable for the proposed occupancy.[9]

We disagree with these recommendations in their present form because, if implemented, they could significantly reduce the ability of local authorities to achieve efficiency in the allocation of resources in the development process. In our discussion of the efficiency of development controls earlier in this chapter, we argued that because of the externalities created by new development, the control of development by local authorities is a possible vehicle by which efficiency can be increased. In particular, it is shown in Frankena and Scheffman

9 Ontario, Planning Act Review Committee (1977), pp. xi–xii.

(1979) that minimum lot-size zoning combined with aggregate development controls is the most plausible policy alternative by which local authorities could enhance efficiency. Furthermore, it is shown in that paper that the lot-size zoning requirements necessary for efficiency would generally differ from one city to another, so that setting provincial standards on lot-size zoning is unlikely to be justified on efficiency grounds.

Of course lot- and house-size zoning may also be used by local authorities to exclude certain types of people from living in particular areas – an issue of equity. As with all policies, there are both efficiency and equity implications of local authorities' development control policies, and we cannot agree that setting provincial guidelines in this case is justified. We believe a more sensible policy would be the provision of a forum through which parties claiming to be inequitably treated can appeal local decisions. Of course we would suggest that both efficiency and equity be given prominence in the adjudication of such appeals.

The PARC Report's recommendation to limit the ability of municipalities to impose financial levies has the same defect as the recommendations proposing lot- and house-size zoning guidelines: it may significantly impair the ability of the local development control process to achieve efficiency. Earlier in the chapter we argued that because of the inefficient incentives created by the property tax system, local development control policies may not attain full efficiency of resource allocation. However, to the extent the allocation is inefficient, as with any inefficient allocation, there is, by definition, a potential for all involved agents to improve their well-being if full efficiency can be attained.

When there are inefficient development controls, the planning authority or city government and the development industry may perceive the mutual advantage to moving to a system of more efficient controls, and an important vehicle for attaining this solution could be the use of financial levies or subsidies by the local authority. For example, to the extent that new development imposes external costs on the existing residents of a locality, on efficiency grounds new-comers should be taxed in excess of the marginal costs of providing services to them and one convenient indirect method of imposing such a tax is to impose financial levies or other requirements on developers. Thus, rather than have the province set guidelines on the requirements municipalities are able to enact, again we would suggest that a more sensible policy would be the provision of a forum for appeal, in which both efficiency and equity implications are fully considered in the adjudication of appeals.

5

Rural and agricultural land use in Ontario

Some of the principal land use issues in Ontario involve rural land, particularly non-farm residential development in rural areas and conversion of good agricultural land to urban and other uses. The political discussion of these issues has been characterized not simply by a lack of economic analysis but by a general ignorance of the nature and the causes of the major rural land use patterns and trends. Consequently, we shall survey the available data on the relevant aspects of rural land use – i.e., rural non-farm residential development and conversion of agricultural land to other uses – to help put the current policy issues in perspective, and thus, we hope, provide a useful background not only for the following chapter dealing with provincial rural and agricultural land use policies but also for public discussion of these issues in the future.

We shall begin by describing briefly the Canada Land Inventory classification of Soil Capability for Agriculture, since most of the current discussion of agricultural land is based on this classification system, and then survey the available data on non-farm residential development in rural areas and conversion of agricultural land to other uses in Ontario, as a background to a discussion of provincial policies aimed at controlling rural severances and preserving agricultural land in the next chapter.

CANADA LAND INVENTORY

The Canada Land Inventory of Soil Capability for Agriculture classifies the mineral soils of the province into seven classes. A lower class number indicates better soil. Table 5.1 shows the relative average yields for field and forage crops that can be expected from the seven classes of land. Soils in classes 1 to 4 are considered by the Canada Land Inventory to be capable of sustained use for cultivated annual field crops, those in classes 5 and 6 only for perennial forage crops, and those in class 7 for neither. The classification does not consider

TABLE 5.1

Performance indices of soil classes (class 1 = 1.00)

Class	Common field crops	Forage crops
1	1.00	1.00
2	0.80	0.80
3	0.64	0.66
4	0.49	0.58
5	n.a.	0.54
6	n.a.	0.44
7	n.a.	n.a.

n.a. = not available
SOURCE: Anderson (1971), p. 40, and Hoffman (1971), p. 38

TABLE 5.2

Acreages of land by class of soil capability for agriculture

Class	Acres ('000)	Per cent
1	4,819	9.1
2	5,273	9.9
3	6,241	11.8
4	5,330	10.1
5	3,395	6.4
6	2,406	4.5
7	19,850	37.4
Organic	5,240	9.9
Unmapped	472	0.9
Total	53,025	100.0

SOURCE: Hoffman and Noble (1975), p. 7

capability of soils for orchards or specialty crops, and organic (non-mineral) soils are not placed in capability classes.

The provincial government normally considers classes 1 to 4 to be 'prime' agricultural land, but others concerned with agricultural land often restrict 'prime' to refer to classes 1 to 3, since class 4 is marginal for sustained arable agriculture. Table 5.2 shows the total acreages and the percentage distribution for the soil classes in the Canada Land Inventory Area of the province.[1] The soils

1 Only the Canada Land Inventory Area, which includes southern Ontario and the southern parts of northern Ontario, is covered by Table 5.2. The total land area of the province is about 220,000,000 acres.

are naturally unevenly distributed, with prime agricultural land (particularly classes 1 and 2) being disproportionately more prevalent in the southwest.

SURVEY OF EVIDENCE ON RURAL AND AGRICULTURAL LAND USE

We have used data available in the *Census of Agriculture* and over a dozen studies of land use in Ontario to try to answer two fundamental factual questions about rural land use:

1. What is the allocation of the various classes of agricultural land among different uses, particularly agricultural and rural non-farm residential uses?
2. How much of the various classes of agricultural land is being permanently and irreversibly converted to non-farm uses, particularly rural non-farm residential and built-up urban uses, each year?

Although these questions may appear simple, the incompleteness and incomparability of the available land use data make it impossible to give anything more than rough approximations as answers. Moreover, even if one had reasonably complete data on land use, it would be difficult to determine which land is being *permanently and irreversibly* converted to non-farm uses (for example, how much of a 20-acre rural non-farm residential lot might be considered unavailable for agriculture in the future). Apart from trying to provide a description of land use, we review the explanations which have been suggested for the observed changes in land use.

Census data

Data from the *Census of Agriculture*, which is carried out by Statistics Canada every five years, reveal that the total acreage of census farms in Ontario decreased between 1951 and 1971 and then remained more or less constant (depending somewhat on whether the 1961 or the 1976 definition of a census farm is used) between 1971 and 1976 (see Table 5.3). The *improved* acreage of census farms also decreased between 1951 and 1971 but then increased somewhat in 1971-6 (using a consistent definition of census farm). The percentage decrease in acreage was greatest in 1966-71, presumably because prices of agricultural products were unusually low relative to prices of farm inputs and the consumer price index. The highly publicized figure of 26 acres per hour pertains to the decrease in improved acreage of census farms in that period. Basically, the data in Table 5.3 indicate 'a continuation of the trend toward contraction of farm acreage that commenced about 1930.'[2]

2 Ontario, Ministry of Agriculture and Food (1972), p. 1.

TABLE 5.3

Area and use of census farms in Ontario, 1951-76 ('000 acres)

	1951 definition		1961 definition				1976 definition	
	1951	1956	1961	1966	1971	1976	1971	1976
Total	20,880	19,880	18,579	17,826	15,963	15,473	14,190	14,744
Improved	12,693	12,572	12,033	12,004	10,865	11,069	9,992	10,708
Unimproved	8,187	7,307	6,546	5,822	5,098	4,404	4,198	4,037

DEFINITIONS

Census farm 1951: A farm of one or more acres with production valued at $250 or more in the previous year, or of three or more acres. 1961: A farm of one or more acres with sales of $50 or more in the previous year. 1976: A farm of one or more acres with sales of $1,200 or more in the previous year.

Improved land Under crops, improved pasture, summerfallow, and other.

Unimproved land Woodland (excluding commercial forests), natural pasture or hay land that has not been cultivated, grazing areas or wasteland, and other.

SOURCE: Statistics Canada, Census of Canada, *Agriculture*, quinquennial

TABLE 5.4

Agricultural production in Ontario, 1951-76 (base 1951 = 100)

Year	(1) Physical volume of agricultural production	(2) Improved farmland	(3) = [(1)/(2)] × 100
1951	100.0	100.0	100.0
1956	105.5	99.0	106.6
1961	131.9	94.8	139.1
1966	155.6	94.6	164.5
1971	168.7	85.6	197.1
1976	178.0	87.2	201.8

SOURCE: Ontario, Ministry of Agriculture and Food, *Agricultural Statistics for Ontario*, annual

Production

In spite of the decrease in the improved area of census farms, aggregate agricultural production increased substantially during the period 1951-76 because of a doubling of output per acre (see Table 5.4).[3]

3 The increase in the index of production per improved acre is due in part to concentration of farming on the more productive acres and changes in the composition of output and hence overstates the average increase in yield for given crops on given parcels of land.

Explanations for the decrease in agricultural land use

Although there are casual studies, there is no careful economic study which explains the changes in agricultural land use in Ontario during the past 25 years. Thus there is considerable uncertainty and disagreement about the relative importance of several possible explanations for the recorded changes in agricultural land use, such as (1) increasing demand for land for built-up urban use, (2) increasing demand for land for rural non-farm residences, and (3) decreasing demand for land for agricultural use (due, for example, to increased productivity in agriculture and an inelastic demand for agricultural products).

One reason for the primitive state of research on this topic is that data on non-agricultural uses of land, land rents, and land prices are very incomplete. Consequently it is difficult even to determine what has happened to the land which has left agricultural production or census farms let alone to explain such changes.

Another problem with attempts to explain land use change is that casual inferences about causation are sometimes based on changes in land use without reference to land rents. However, elementary economic analysis indicates that a given change in land use could be explained in more than one way; for example, land might be converted from agricultural to non-agricultural use because of a decline in demand on the part of agriculture or an increase in demand on the part of non-agricultural uses or a variety of combinations of shifts in the two demands. There is thus wide scope for incorrect inferences in studies which use data on land use without data on land rents. As we discuss further below, there is also scope for confusion between the role of land rents and land prices in explanations of land use change.

(a) Changes in the economics of farming

Some observers appear to believe that the most important explanation for the decrease in the total and improved areas of census farms involves changes in the economics of farming which are unrelated to increases in competing demands for land from non-farm uses. For example, in a study of changes in the numbers and acreages of farms in regions of Canada based on different Census Metropolitan Areas between 1961 and 1971, Bryant concluded that by far the largest proportionate decreases in farm numbers and acreages occurred in the Maritimes, northern Ontario, and Quebec, which were the areas with the smallest urban population increases and the poorest agricultural land. He concluded that 'this must be explained by the "uncompetitive" nature of the farmland in these regions rather than metropolitan influences ... [In these areas] removal of land and labour from agriculture has occurred very rapidly during the 1960's and

1970's, and it is suggested that this relates to the marginality or uncompetitiveness of agriculture in these regions.'[4]

As another more specific example, it was recently reported that 'high yielding tobacco varieties have meant less land is needed to meet demand ... With the tobacco acreage on sandy soils in Southwestern Ontario diminishing, 60,000 to 70,000 acres are freed for alternate crops.'[5] Translated into the conventional Ricardian model familiar to economists, this explanation can be demonstrated with the help of Figure 5.1. For the sake of simplicity we portray only two of the qualities of agricultural land, class 1 and class 4. The initial non-land unit costs of production for tobacco per acre of land for the two classes are given by curves AC_1 and AC_4. The industry supply and demand curves are given by S and D, the price and total output of tobacco are P and Q, the competitive Ricardian rents per acre of tobacco land for the two classes are the shaded areas R_1 and R_4 respectively, and output per acre for the two classes is given by q_1 and q_4 respectively. As a result of technological improvements, suppose the cost curves shift down to AC_1' and AC_4', in such a way that the point of minimum average cost of production shifts substantially to the right and the supply curve shifts to S'. The price of tobacco falls to P'. The total output of tobacco rises to Q', but assuming a reasonably inelastic demand curve for tobacco the percentage increase in total output $(Q' - Q) / Q$ is less than the percentage increase in output on class 1 land $(q_1' - q_1) / q_1$. The implication is that less land is used for tobacco, output of tobacco on class 4 land ceases, and the Ricardian rent on class 4 land drops to zero.[6]

The question which remains is how important such decreases in demand for land for agricultural use have been compared to increases in competing demands. Most observers appear to believe that increases in two types of competing demands have played a role in reducing agricultural land use. The first is urban demand for land for housing, firms, and related uses. The second is non-farm demand for land in rural areas, such as for rural residences.

4 Bryant (1976), pp. 69, 71.

5 *London Free Press*, 25 June 1977, p. 3.

6 The model might be more realistic if it took account of the demand for land for non-agricultural uses. If this were included, technological improvements in production of tobacco might lead to conversion of class 4 land to non-farm use rather than to its abandonment. Also, even if the current rent on class 4 land were zero in both agriculture and non-farm use, as long as rents were expected to be positive in either use at some time in the future the land would command a positive price and simply be left idle rather than being literally abandoned.

Figure 5.1
Ricardian model of tobacco farm land abandonment

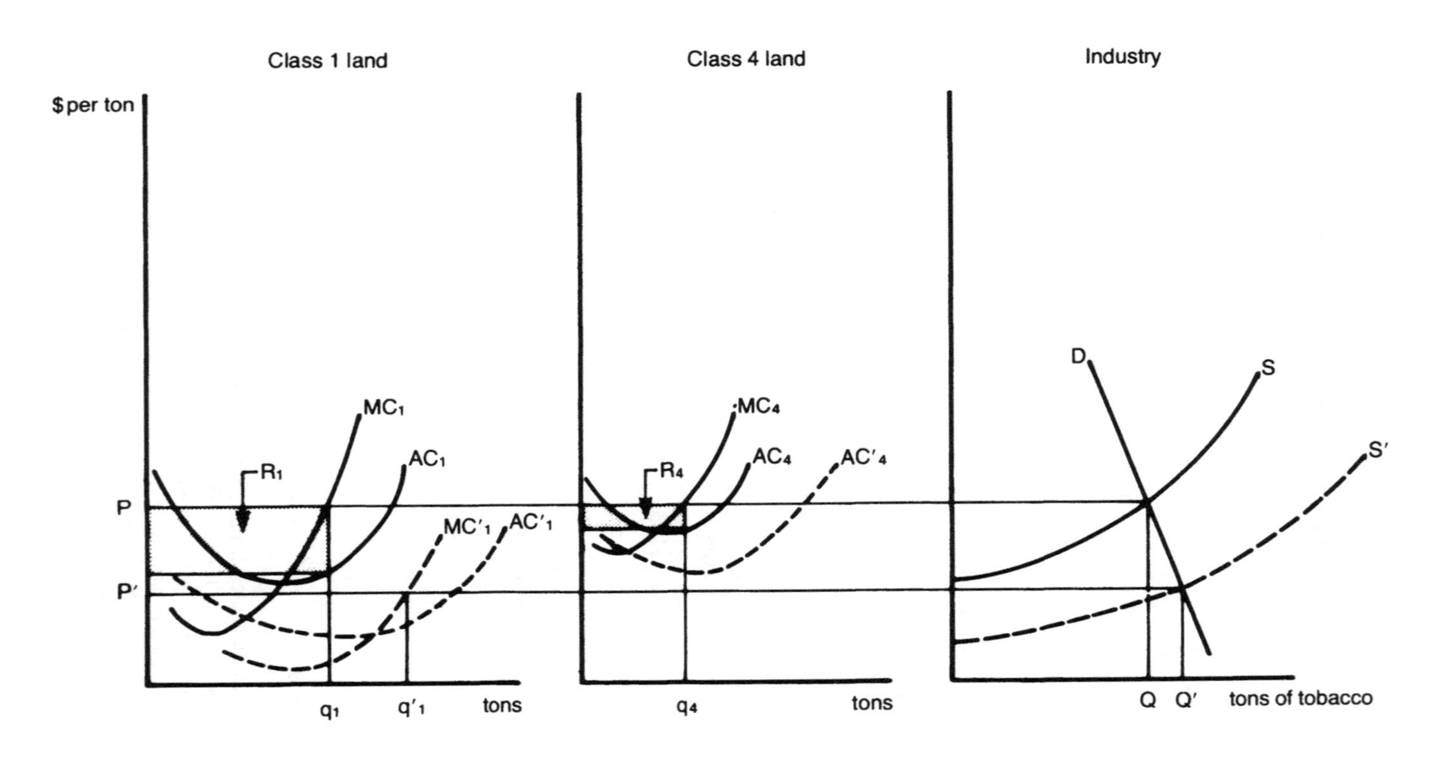

(b) Urban land use

Studies of rural land use change have sometimes assumed that the decline in farmland, or at least the decrease in the improved area of census farms, is due entirely to expansion of urban land use. For example, in a study for south-western Ontario between 1941 and 1961, Russwurm states that 'the loss of agricultural land ... is usually stressed as a summation effect of urbanization on agriculture. The loss of improved farmland is used to measure this summation effect. Since abandonment of marginal land is negligible in the study area the loss of improved farmland is almost solely attributable to conversion into urban land uses.'[7]

In spite of such assertions, in the aggregate it seems clear that increased urban demand is in fact responsible for at most a minor share of the decrease in agricultural acreage. During the period 1951-76, the area of census farms decreased by about five million acres (see Table 5.3). The most reliable estimates suggest that the area devoted to built-up urban land uses increased by about 10 per cent of this,[8] and some of the expansion of urban uses occurred on land which was not previously part of census farms. The rate of conversion of land to built-up urban use in Ontario is about 20,000 acres per year, 63 per cent coming from former improved agricultural land and 79 per cent from former prime (classes 1 to 3) agricultural land. Thus, about one per cent of the stock of good agricultural land was converted to built-up urban use during the past *decade*. This loss is quite low compared to the increase in productivity per acre in agriculture during the same period. Also, conversion of land from agricultural to urban use might have been due to a decline in demand for land for agricultural use as well as an increase in demand for land for urban use.

(c) Rural non-farm uses

It has now become common to blame increasing demand for rural non-farm uses, particularly residences, for the decline in agricultural land use (and/or for the irreversible removal of land from the stock available for agricultural production in the future). For example, the Bureau of Municipal Research asserts that 'the weight of expert opinion is that non-farm uses of rural land now are the predominant influence in farm land withdrawal.'[9] Referring to 'the rapid loss of non-marginal improved land, particularly during the 1966-1971 period, 'Rodd and Van Vuuren state that 'it is highly unlikely that all this land was rejected

7 Russwurm (1967).

8 The evidence referred to here and in the remainder of this section is based on the studies which are discussed in detail in the final section of the chapter.

9 Bureau of Municipal Research (1977), p. 4.

from agriculture because it could not yield a reasonable return in farming. It is more likely that non-farm demands were so fierce that a higher return than from farming could be obtained. Without this competition much of this land would have remained in agriculture and would have been profitably farmed ... It appears that the vast majority of this land retiring from agriculture was neither used for city expansion nor sub-marginal. There is evidence that one of the greatest causes in this retirement process is the rapid proliferation of rural non-farm housing.'[10]

On the basis of available studies, it would be difficult to support or refute such assertions conclusively, but the evidence appears to be consistent with the view that changes in the acreage of census farms have resulted at least as much from other changes in the economics of farming as from increases in the demand for rural non-farm residences.

It is difficult to quantify the rate of conversion of land from agricultural to rural non-farm residential use,[11] but a number of significant statements can be made about rural non-farm residential land use. First, in much of rural southern Ontario it appears that the density of non-farm residences is between one and five per square mile and that one or two per cent of the land in good agricultural areas is actively used for non-farm residential purposes (excluding hamlets and villages), although several times this much land is owned by resident non-farmers. Second, the rural non-farm population in areas within commuting distance of major cities in southern Ontario roughly tripled between 1951 and 1971, the rate of increase being greater before the mid-1960s than later. Third, rural non-farm residences have absorbed a disproportionately larger share of lower-quality agricultural land than prime agricultural land. These findings cast doubt on the accuracy of the more extreme statements that have been made about the importance of rural non-farm residential development in explaining recent trends in agricultural land use.

(d) Speculation and rising land prices

There have been major changes in ownership of land in rural Ontario. Most significant is the fact that ownership of farmland by farmers has decreased considerably. Because of this change, some observers believe that a significant amount of land has been removed from agricultural production and census farms

10 Rodd and Van Vuuren (1975), pp. 110–11. The authors do not present the evidence they claim to have in the last sentence quoted.

11 It is frequently argued that, although the percentage of rural land used for non-farm residences is generally low, scattered non-farm rural residences have severe adverse effects on agriculture in surrounding areas. See James F. MacLaren, Ltd. (1975) and Rodd (1976a). These arguments are considered in Chapter 6.

by 'speculators' and others who hold land for capital gains. This argument has little appeal a priori, however, since it assumes that speculators irrationally hold their land idle rather than renting it to farmers.[12] It is also contradicted by the empirical evidence. Apart from land converted to built-up urban use and small rural severances, there is no evidence that the sale of farmland by farmers to non-farmers has influenced land use to a significant extent, since much of the farmland purchased by non-farmers is rented and continues to be farmed by the previous owner or other tenants, and much of the farmland which is not farmed after the ownership change was previously classified as unimproved farmland and would not be farmed anyway.

Some observers also suggest that significant amounts of land have been removed from agricultural production because of rising land prices brought about by the expectation that the land will beused for urban or non-farm residential purposes in the future. However, this reasoning confuses the role of land *rents* (per time period) and land *prices*. During any given time period, land will be allocated to the use which is willing to bid the highest *rent* for it. The price of a parcel of land is normally equal to the present discounted value of the expected rents. When the expected future non-farm demand for land increases so that expected non-farm rents exceed farm rents beginning at some date in the future, the price of the land in question will naturally rise above the level determined by the present discounted value of agricultural rents. However, this does not change supply and demand for use of the land in the present, and hence neither the present rent nor the present land use would normally be affected. Of course, the incentive to make agricultural improvements in the land might be reduced, but typically this would not bring an end to agricultural production.[13]

STUDIES OF RURAL LAND USE

We shall now review in some detail the findings of over a dozen studies of rural land use in Ontario. The conclusions presented above concerning the causes of the decline in agricultural land use were reached on the basis of the information in these studies. The detailed information in the present section will be useful to

12 By renting otherwise idle land to a farmer, the owner would not only earn rent from the farmer but would qualify for two significant provincial subsidies for agricultural use of land: (1) properties used as farms are assessed by the province at a much lower percentage of market value than are non-farm properties, and (2) farms receive rebates from the province of 50 per cent of the property taxes paid.

13 In fact, if there were market value assessment for non-farm properties and lower assessments and/or rebate of property taxes on farms, there would be an increase in the incentive to continue agricultural production if the land price increased.

TABLE 5.5

Percentage distribution of land among active uses in Punter's study area, 1954 and 1971

Active land use	1954	1971
Residential	1.4	3.1
Farm	0.5	0.2
Non-farm	0.9	3.0
Recreational	0.3	2.2
Agricultural	49.7	34.6
Owner-occupied	39.8	12.6
Other	9.9	22.0
Other	0.6	2.7
Total	52.0	42.6

NOTE: Land not considered in active use included idle land owned by farmers, speculators, developers, etc., and conservation areas not used for recreation. See Punter (1974), pp. 61-8

SOURCE: Punter (1974), p. 318, and Table 23

anyone involved in the current debate over provincial rural and agricultural land use policies.

Punter

Punter (1974) provides evidence on land use and ownership between 1954 and 1971 in four rural townships in the Toronto-centred region. Table 5.5 reproduces some of Punter's data on land use. These data reveal that even in 1954 only a little over half of the land in the study area was 'actively' used. By 1954 large amounts of marginal agricultural land had already been dropped from cultivation for reasons which had nothing to do with non-farm demand for land. The Ricardian rent on this lower-quality agricultural land had simply dropped to zero.

Between 1954 and 1971 the acreage of active agricultural land continued to decrease, from 49.7 to 34.6 per cent of the total land in the study area. Simultaneously, there was a tripling of active non-farm residential land use. However, Punter concludes that the increase in non-farm residential land use was not important in explaining the decline in agricultural land use. For one thing, only 3 per cent of the land was in active non-farm residential use in 1971. Also, according to Punter (p. 26), 'the areas apparently favoured by exurbanities are also very poor areas agriculturally – the thin soils of the escarpment and sandy soils of the moraine have never proved very suitable for productive agriculture ... Exurbanities seem to prefer those areas where the landscape is best which is, conversely, very often where the soils are worst.'

There were also substantial changes in the ownership of land in Punter's rural townships. There was a very large drop in the amount of land owned by farmers as farms were sold to non-farmers, and ownership of land by both non-residents and non-farm residents increased greatly. The percentage of land in owner-occupied non-farm properties doubled between 1954 and 1971 to about 22 per cent of total land.[14] However, Punter concludes that changes in ownership per se had little effect on agricultural land use since non-farmers rented out most of their usable farmland to farmers.

An extensive selection of quotations from Punter (1974) is relevant to these points.

The main impact of exurbanisation on agriculture has been to diversify its character and ownership structure rather than to reduce its actual extent. The data does not support the commonly held view that absentee or exurbanite ownership has been responsible for a major decline in agricultural activity in these areas. [p. 319]

To be sure some 15 per cent of the land has passed out of agricultural use ... This decrease is clearly significant, but it would be wrong to attribute it solely to exurbanisation. Large scale abandonment of agricultural land was already underway prior to the main era of exurban development ... The evidence suggests that the decline in agricultural activity would have been just as great without the exurban influx ... The evidence suggests that the greater part of the agricultural land which is not owned by bona fide farmers is farmed as seriously and productively as that which is ... There has not been a large increase in the amount of 'idle' land. Perhaps some 10 per cent of these areas has been rendered idle by abandonment of agriculture, and changes in ownership. [pp. 379-81]

In contrast, Punter also found that about 26 per cent of the 'idle' land in 1971 belonged to rural residential lots (37 per cent if one includes part-time residential), while another 6 per cent consisted of vacant severed rural lots on which houses had not yet been constructed.[15] Punter's discussion suggests that some of this idle land was in good agricultural areas such as Whitby Township and would have had a positive rent in agricultural use in 1971 and/or possibly at some date in the future if it had not been divided into 10-acre lots and similar parcels.[16]

14 Punter (1974), pp. 126, 295.
15 Ibid., p. 336.
16 Ibid., p. 393.

Punter suggests that if planning controls had not imposed a 10-acre minimum for severances for a number of years, many people would have opted for smaller lots of around two acres and less agricultural land would have been severed per residence.[17]

Punter also presents some information on the total acreage of 10- and 25-acre lots in 13 townships around Toronto in 1967: 'By 1967 ten and twenty-five acre severanges occupied large acreages in all the townships that constitute the main exurban belt around Toronto ... In the thirteen townships for which data is available over 4000 properties of 10-25 acres had been created constituting approximately 9 percent of the total area of these townships ... As a general rule a maximum of 20-30 percent of this area would be occupied by 25 acre parcels ... Over 60 percent remained vacant in 1967 but there was nothing that could be done to prevent any owner from obtaining a building permit for a single family home' (p. 169).

Martin

Martin (1975) found that in a large rural-urban fringe area northeast of Metro Toronto between 1961 and 1971 cropland, cultivated pasture, and related land uses decreased from 68.3 to 64.6 per cent of the total land area while rough pasture, grazing land, and scrub woodland decreased from 18.6 to 16.1 per cent (see Table 5.6). During the same period, non-farm residential use increased from 1.6 to 3.1 per cent, recreational use increased from 1.0 to 1.9 per cent, extractive use increased from 1.3 to 3.0 per cent, other 'urban' uses (industrial, etc.) increased from 0.3 to 0.7 per cent, and woodland increased from 8.8 to 10.5 per cent.

These statistics and Martin's other findings indicate several important points. First, in spite of a substantial reduction in the number of owner-operated farms during this period, the amount of 'active' farmland decreased only moderately. Thus change of ownership per se had little effect on the use of high-quality agricultural land. Hobby farmers, speculators, and developers who bought improved farmland typically rented it back to the former owners or leased it to neighbouring farmers.

Second, even in 1971 only about three per cent of the land was used for non-farm residences. This figure includes several towns (Markham, Stouffville, Uxbridge) as well as smaller villages and hamlets; thus only about one per cent of the land outside of towns, villages, and hamlets was used for non-farm residences

17 Ibid., pp. 337, 381. Of course the aggregate effect of planning controls on total acreage severed would also depend on the effect they had on the number of lots severed.

TABLE 5.6

Land use in Martin's study area in the rural-urban fringe northeast of Toronto, 1961 and 1971

	Per cent of land	
Land use	1961	1971
DECLINING USES		
Cropland, cultivated pasture, orchards, vineyards, small fruit and vegetable fields, nurseries	68.3	64.6
Rough pasture, grazing land, scrub woodland	18.6	16.1
INCREASING USES		
Non-farm residential[a]	1.6	3.1
Public and private recreation	1.0	1.9
Extractive	1.3	3.0
Industrial, warehouse, fuel storage, transport, communications, utilities, schools, churches	0.3	0.7
Woodland	8.8	10.5

a Includes towns, villages, and hamlets, scattered built-up residential areas, and scattered isolated residences. Scattered isolated residences with lots under one-half acre were assumed to have an average size of 0.3 acre. Scattered isolated residences with lots between one-half and five acres were assumed to have an average size of three acres.
SOURCE: Martin (1975), Tables 16-17, Maps 8, 10

in 1971. The land absorbed by non-farm residences (excluding scattered isolated residences) between 1961 and 1971 came almost entirely from the two categories of declining land use, cropland and cultivated pasture on the one hand and rough pasture, grazing land, and scrub woodland on the other, nearly in the same proportions that these two categories of land existed in the study area.

The data on non-farm residences do not include hobby farms. Martin found that hobby farms tended to be concentrated on the least agriculturally productive but most scenic land in the study area, particularly in the Oak Ridge Moraine.

Rodd

Rodd (1975, 1976b) studied land use in the early 1970s in a random sample of 18 of the 102 townships in the area lying approximately 25 to 75 miles from Toronto. His survey was restricted to rural properties and excluded the contiguous built-up areas of cities, towns, villages, and hamlets, and also transportation rights-of-way. Rodd presents data on the percentage distribution of properties (not acres) among major uses (farms, houses, estates, etc.) in 1973

TABLE 5.7

Rural land use in Rodd's sample townships

Type of property	(1) Percentage of properties, 1973	(2) Average acreage of properties sold, 1969-72	(3) Estimated percentage of acreage, 1973
Farms with buildings	32.4	97.7	80.7
Farms without buildings	4.0	43.6	4.4
Estate	8.6	10.2	2.2
House	40.0	4.2[a]	4.3
Cottage	2.6	4.2[a]	0.3
Vacant	5.6	21.2	3.0
All other	6.7	29.4	5.0

a Average acreage was computed for houses and cottages together.
SOURCE: Rodd (1976b)

TABLE 5.8

Lot size and soil capability for agriculture for houses and estates sold in 1969-72 (number of lots)

Lot size (acres)	Soil capability for agriculture 1	2	3	4	5	6	7	Org.	Total	Per cent
0–1	126	64	21	70	7	3	3	15	309	57.8
1–3	20	15	3	26	3	3	1	2	73	13.6
3–5	12	12	8	25	5	0	2	4	68	12.7
5–10	15	3	5	32	1	0	0	2	59	11.0
10–20	0	1	1	2	0	0	0	1	7	1.3
⩾20	3	1	2	6	2	0	0	1	17	3.2
Total	176	96	41	162	19	6	6	28	535	100.0
Per cent	32.9	17.9	7.7	30.2	3.6	1.1	1.1	5.2	100.0	100.0

SOURCE: Rodd (1975), p. 55

(Table 5.7, column 1). He also presents data on the average acreage of properties which were sold during 1969-72 in a number of categories (column 2). If we assume that the properties sold were representative of all properties in the same category in terms of acreage, we arrive at the distribution of acreages presented in column 3.

The date in Table 5.7 indicate that estates, houses, and cottages combined accounted for just under 7 per cent of rural land. Table 5.8 presents a cross-tabulation of data on the distribution by lot size and soil capability for agriculture for houses and estates sold in 1969-72. Several interesting points emerge from these data:[18]

1. Although lots smaller than one acre accounted for 58 per cent of all lots by number, they accounted for only 10 per cent of all lots by acreage. By contrast, although lots of 10 or more acres accounted for 4.5 per cent of all lots by number, they accounted for 35 per cent of all lots by acreage.
2. Although 51 per cent of all non-farm residential lots were on soils in classes 1 and 2 in terms of capability for agriculture, these lots accounted for only 35 per cent of the area of all non-farm residential lots because the average size of lots on the best agricultural land was smaller than the average for all lots. Since about 48 per cent of the land (excluding built-up areas) in the townships in the sample was in soil classes 1 and 2, it can be concluded that on the basis of acreage *rural non-farm residences were located disproportionately on lower-quality agricultural soils.*[19]
3. Thirty per cent of all residential lots, accounting for 42 per cent of the area of all residential lots, were on soils in class 4 in terms of capability for agriculture. By comparison, only about 11 per cent of the land in the townships in question was in soil class 4.

It is natural to compare the findings in the studies by Punter, Martin, and Rodd for land ownership and use in the rural townships of the Toronto-centred region around 1971. Punter found that 22 per cent of the land in his rural townships was in non-farm owner-occupied properties while Rodd's data suggest that 7 per cent was in estates and houses. Evidently there are two major explanations for the difference in these figures. First, Rodd selected a random sample of townships whereas Punter deliberately selected a major portion of his sample from areas with high non-farm ownership, particularly King Township. Second, Rodd did not count hobby farms as estates and houses while Punter evidently included hobby farms owned by non-farmers in with non-farm properties. Based on the two studies, figures of 15 and 8 per cent respectively would appear to be reasonable though crude guesses for the shares of land in the Toronto-centred region in owner-occupied properties owned by non-farmers, including and excluding hobby farms respectively.

18 Acreage calculations assume that the mean lot size in each size category was equal to the mid-point of the category and that the mean size for lots of 20 or more acres was 25 acres.

19 The percentage distribution of total land area for the townships among soil classes was based on data in Hoffman and Noble (1975) for 14 of the 18 townships in the sample. Data for the other 4 townships were not available.

Punter and Martin both found that about three per cent of the land in their study areas was in 'active' non-farm residential use (as opposed to ownership). However, Martin's study area included several towns as well as villages and hamlets while Punter's area included only a few minor villages. Thus, considering only areas outside towns, villages, and hamlets, Punter and Martin found that about three and one per cent respectively of land was in active non-farm residential use. On the basis of the two studies, a figure of about two per cent would appear to be a reasonable though crude guess for the proportion of rural land in active non-farm residential use within commuting distance of Toronto in 1971.[20]

Michie and Found

Michie and Found (1976) have studied the distribution of hobby farms (which they call 'rural estates') in the Toronto-centred region in 1969-72. They found that hobby farms are very heavily concentrated on the Oak Ridge Moraine and the Niagara Escarpment, where the land has a low capability for agriculture. They note that 'although agriculture is still common throughout much of the regions, abandonment of cultivation, and even grazing, has been occurring throughout this century ... Low land capability for agriculture has led to low land prices, which has facilitated the invasion of the regions by non-farmers whose primary income source lies in the city.'[21]

Krueger

Krueger (1978b) used air photos and field checks to count the number of 'urban' residences in each township concession block in the Niagara fruit belt in 1934, 1954, 1965, and 1975 (see Table 5.9). The data in Table 5.9 reveal a continual decrease in the proportion of blocks with 0 to 2 urban residences and a continual increase in the proportion of blocks with 25 or more urban residences. In 1975 the average density of scattered non-farm residences in rural townships (i.e., townships without contiguous developed areas covering at least 12 acres and having a density of two or more buildings per acre) was about 10 per square mile.

Avey

Avey (1974) studied the distribution of non-farm residences in 1973 in a 540-square-mile region immediately northwest of London. He found that the average

20 Assuming a grid road system with roads spaced at one-mile intervals, and assuming an average of two acres of land in active residential use per property, this figure would imply three non-farm properties per mile of rural road.

21 Michie and Found (1976), pp. 17, 20.

TABLE 5.9

Urbanization in the Niagara Fruit Belt[a] (per cent of township concession blocks)

Number of urban residences[b] per township concession block[c]	1934	1954	1965	1975
0–2	59	46	39	28
3–6	19	16	19	17
7–24	17	30	30	21
⩾25	5	8	12	34
Total	100	100	100	100

a The Niagara Fruit Belt is defined as the townships of Barton, Saltfleet, North Grimsby, Clinton, Louth, Pelham, Grantham, Thorold, Niagara, and Stamford.

b Defined as homesteads counted on air photos minus farmsteads determined by field checking.

c Township concession blocks are of roughly but not exactly equal size and average around 0.4 square mile in area.

SOURCE: Krueger (1978b), Table 2

density of non-farm residences decreased with distance from the city. Average density was about five non-farm residences per square mile one mile from London, about three per square mile 10 miles from London, and very roughly one per square mile beyond 20 miles from London.[22]

These figures suggest that apart from areas within commuting distance of Toronto, Ottawa, and Hamilton, on average less than one per cent of the land in good agricultural areas in rural Ontario is actively used for non-farm residential purposes.

Russwurm

Russwurm (1976) has used 1951, 1961, and 1971 Census data to estimate the population living within 25 to 30 miles of the centre of London in each of the following residential categories: (a) built-up area of London; (b) built-up area of surrounding towns, villages, and hamlets with populations over 50; (c) scattered non-farm residential; and (d) farm. Since the Census definition of farm is used, people living on hobby farms who rent out land for farming purposes would be classified in the farm population. He has done the same thing for the population living in non-overlapping areas within roughly 25 miles of St Thomas, Chatham-Wallaceburg, Sarnia, Stratford, and Woodstock. Russwurm's data are

22 Avey (1974), p. 80, Maps 2, 7, Figure 3.

TABLE 5.10

Population in Russwurm's southwestern Ontario regional cities

	Population ('000)			Per cent of total population		
	1951	1961	1971	1951	1961	1971
Urban	331	441	542	70	74	78
Scattered rural non-farm	25	60	77	5	10	11
Farm	122	96	80	26	16	12
Total	478	597	699	100	100	100

SOURCE: Russwurm (1976)

summarized in Table 5.10, where categories (a) and (b) have been combined into 'urban' and data for all five areas have been aggregated.

One finds that urban and rural non-farm populations have increased while farm populations have decreased in each of these areas. The scattered rural non-farm population tripled between 1951 and 1971. One also finds that the scattered rural non-farm population has increased faster than the urban population in percentage terms, but that the absolute as well as relative growth of the scattered rural non-farm population was less in 1961-71 than in the preceding decade. Russwurm states that 'personal knowledge and field experience confirms that increasing planning controls along with a slowdown in growth rate are causal factors.'[23]

In another study, Russwurm (1974) reports that 'despite much greater difficulty in getting land severances approved and increasingly stronger planning controls, the number of residences of urbanites grows in the countryside around all cities' (p. 173). Russwurm found that in the six-mile-wide band around the 1973 built-up area of Kitchener-Waterloo the number of residences increased by 88 per cent between 1961 and 1973, from an average of a little over two to an average of a little over four residences per square mile.

Gierman (1977)

Gierman (1977) provides measures of the conversion of land from rural to built-up urban use in all 24 urban areas in Ontario which had a population of 25,000 or over in 1971. He provides not only total acreage (Table 5.11) but also

23 Russwurm (1976), p. 85.

TABLE 5.11

Built-up areas of major urban areas in Ontario, 1966 and 1971

	Area (acres)		
Urban area[a]	1966	1971	Increase 1966-71
Barrie	6,909	7,666	757
Belleville	3,652	4,223	571
Brantford	10,162	11,641	1,479
Chatham	4,306	4,816	510
Cornwall	4,801	4,958	157
Guelph	6,787	8,000	1,213
Hamilton	38,197	43,251	5,054
Kingston	11,260	13,637	2,377
Kitchener	19,769	26,018	6,249
London	26,911	30,730	3,819
North Bay	8,687	9,314	627
Oshawa	18,024	19,528	1,504
Ottawa[b]	66,335	77,740	11,405
Peterborough	10,413	12,067	1,654
Sarnia	12,265	13,687	1,422
Sault Ste Marie	11,713	12,860	1,147
St Catharines–Niagara	41,285	53,857	12,572
Sudbury	20,779	23,761	2,982
Thunder Bay	14,944	16,362	1,418
Timmins	3,575	3,766	191
Toronto	185,760	214,807	29,047
Trenton	4,087	4,735	648
Windsor	27,545	30,572	3,027
Woodstock	5,831	6,660	829
Total[c]	563,997	654,656	90,659

a The urban area is all land in a CMA, CA, or other urban centre of over 25,000 population that is classified as 'built-up' according to the Canada Land Inventory Land Use Classification. Built-up areas include 'all compact settlements, i.e., the built-up parts of cities, towns and villages, including any non-agricultural open space that forms an integral part of the urban agglomeration, such as vacant lots, town parks, etc. ... Some farmsteads and areas of low-density urban sprawl are also included, especially strip development along roads ... This category also includes isolated units which are separated from compact settlements and are used for industrial, commercial and associated urban purposes ... Major highways (4 lanes) and major interchanges are included in built-up area.'

b Includes Hull, Quebec

c Includes only the 24 urban areas listed

SOURCE: Gierman (1977), Table 2

TABLE 5.12

Former land use of rural land converted to urban uses in Ontario, 1966-71

Farm use	Acres	Per cent
Improved agricultural land	56,860	62.72
Cropland, improved pasture	49,716	54.84
Horticulture	1,855	2.05
Orchards, vineyards	5,289	5.83
Unimproved pasture	18,454	20.35
Forest land	13,261	14.63
Unproductive	5,200	5.74
Productive	8,061	8.89
Swamp, marsh	500	0.55
Barren land	632	0.70
Extractive	952	1.05
Total	90,659	100.00

NOTES: See Table 5.11
SOURCE: Gierman (1977), Table 3

TABLE 5.13

Agricultural capability of rural land converted to urban uses in Ontario, 1966-71

Agricultural class	Acres	Per cent
Class 1	30,841	34.01
Class 2	22,098	24.38
Class 3	18,564	20.48
Class 4	5,074	5.60
Class 5	3,623	4.00
Class 6	2,712	2.99
Class 7	3,943	4.35
Organic soil	968	1.07
Unclassified	2,836	3.12
Total	90,659	100.00

NOTES: See Table 5.11
SOURCE: Gierman (1977), Table 5

acreage broken down by former land use (Table 5.12) and by agricultural capability (Table 5.13). Because this is the most detailed available study of rural-urban land conversion throughout the province, we have relied heavily on it in reaching our conclusions in the previous section.

TABLE 5.14

Percentage distribution of land not in built-up urban use in Gierman's study area in 1964

Use	1964	1973
Built-up urban	0.0	7.8
Improved pasture and forage crops	47.0	38.5
Unimproved pasture, range land, non-productive woodland	24.5	23.2
Productive woodland	18.6	20.2
Other	9.9	10.3
Total	100.0	100.0

SOURCE: Gierman (1976), Tables 1 and 3

Gierman found that in these 24 urban areas the average annual rate of rural-urban land conversion in 1966-71 was 18,132 acres, of which 11,372 acres (63 per cent) was formerly improved agricultural land and 14,301 acres (79 per cent) was formerly prime (class 1 to 3) agricultural land.

In the case of improved agricultural land, the 56,860 acres converted to urban use in 1966-71 represent 0.5 per cent of the 1966 stock in census farms and 5 per cent of the decrease in the stock in census farms during 1966-71. In the case of prime (class 1 to 3) agricultural land, the 71,503 acres converted to urban use in 1966-71 represent 0.4 per cent of the 1966 provincial stock.

Gierman (1976)

The study by Gierman (1976) of land use in the Ottawa-Hull region between 1964 and 1973 deals with an area measuring roughly 16 by 23 miles and centring on Ottawa. Because all parts of the study area lie within commuting distance of Ottawa, it is not surprising that Gierman found that the conversion of land from non-urban to urban use was the major land use change in the area.

Gierman found that in 1964 about 201,000 acres of land in the study area were not in built-up urban areas. Table 5.14 shows the percentage distribution of this land in 1964 and 1973. It can be seen that the proportion in improved pasture and forage crops (the only major improved agricultural land use in the study area) decreased from 47.0 to 38.5 per cent of the land which was not in built-up urban use in 1964. At the same time, 7.8 per cent of the land which was not in built-up urban use in 1964 was so used in 1973. Although the decrease in acreage of improved pasture was almost equal to the increase in acreage of built-up urban areas, other data presented by Gierman show that the transfer

TABLE 5.15

Percentage distribution of land by capability for agriculture in Gierman's study area

	Non-urban land, 1964	Net decline in non-urban land, 1964-73
Classes 1 to 3	48.9	52.5
Class 4	13.1	11.9
Classes 5 to 7	33.3	30.3
Other	7.7	5.1

SOURCE: Gierman (1976), Tables 16a and 17a

was not direct. Only about half of the land which became urban during 1964-73 was classified as improved agricultural land in 1964.

Gierman also presents data on the agricultural capability of land which was not in 'urban' use (i.e., not in built-up urban areas, in extractive use, or used for outdoor recration) in 1964 and on the net decline in non-urban land between 1964 and 1973 (see Table 5.15). He found that prime agricultural land in classes 1 to 3 of the Canada Land Inventory accounted for a somewhat larger share of the net decline in non-urban land during 1964-73 (52.5 per cent) than of the initial stock of non-urban land in 1964 (48.9 per cent). In other words, to some extent urban growth was disproportionately concentrated on prime agricultural land.

Special Committee on Farm Income in Ontario

The 1969 report of the Special Committee on Farm Income in Ontario predicted that, of the approximately 17.6 million acres of land in farms in 1966, by 1981 about one per cent (155,000 acres) would be converted to urban use, another one per cent (174,000 acres) could be converted to urban oriented uses, and about one-half per cent (86,000 acres) would be converted to highways, railways, power corridors, and recreation and conservation areas.

University of Guelph Centre for Resources Development

A 1972 ARDA report prepared by the University of Guelph Centre for Resources Development[24] predicted that between 1971 and 1991 under the most likely assumptions 172,000 acres of land would be developed for urban use in southern Ontario and 136,000 acres would be 'idled but not actively developed' as a result of the urban development process. It was estimated that about 60 per cent of all this land would be class 1 and 2 agricultural land. Thus, over the 20-year period

24 Ontario, Ministry of Agriculture and Food (1972).

in question, it was expected that about one per cent of all class 1 and 2 agricultural land in the province would be converted to built-up urban use and another 0.8 per cent would be 'idled' by the urban development process.[25]

The same report includes a study of the number of low-density rural residential Ontario Hydro customers in 1960 and 1970. The study concludes that 'the distribution of low residential density hydro customers in 1960 displays a distinct urban gradient being concentrated particularly around the larger cities ... If we assume that an average customer in this category consumed 5 acres of land the loss of land from farming during the period could be of the order of 1 to 2 acres per 100 acres in some of the townships bordering the urban areas. In other areas the density of customers is not so great, and if we assume an average density today of 0.7 per 100 acres for all of southern Ontario, at maximum some 3% of the area of southern Ontario occupied by farms in 1951 would have been lost to low density residential areas between 1960 and 1970.'[26]

Ministry of Agriculture and Food

In *A Strategy for Ontario Farmland*, issued in 1976, the Ministry of Agriculture and Food estimated that during the period 1976-2000 between 256,000 and 370,000 acres of land would be needed for built-up urban uses (under the implicit assumption that the entire population increase in the province would take place in urban areas) and for highways and utility lines. This comes to about 10,000 to 15,000 acres per year. It should be noted that this is lower than any of the other predictions concerning the amount of land that will be used for urban-related purposes in the coming decades in Ontario.

Peter Barnard Associates

Peter Barnard Associates has estimated that in the period 1976-2000 about 278,000 acres of land will be needed for housing in Ontario. According to the report, this would imply a total urban land requirement of around 500,000 acres, or 20,000 acres per year.[27]

Bird and Hale and M.M. Dillon

An Urban Development Institute study by Bird and Hale, Ltd., and M.M. Dillon, Ltd. (1977) estimated from air photos that 520,000 acres of land in Ontario were developed for non-agricultural uses between 1951 and 1971. However, the report does not give any breakdown of the acreage among the non-agricultural

25 Ibid., pp. 78–93.
26 Ibid., pp. 103, 102.
27 Ontario, Ministry of Housing (1977a).

uses involved or among agricultural soil classes, and the estimate is subject to an unusually wide margin of error.

Summary

The 14 studies reviewed in this section are summarized at some length above (see pp. 66-71). The data on land use conversion indicate clearly that in the aggregate the rate of conversion of land to built-up urban use is low in relation to the rate of productivity increase in agriculture, the stock of agricultural land, and the decrease in the acreage of census farms. During the past decade about one per cent of the good agricultural land in the province was converted to built-up urban use, and rural-urban land conversion amounted to about 10 per cent of the decrease in the area of census farms.

The data also indicate that in the aggregate the rate of conversion of land to active rural non-farm residential use has been even lower than the rate of conversion to built-up urban use. Furthermore, the land converted to non-farm residential use has come in disproportionately larger amounts from lower-quality agricultural land than from prime agricultural land. This finding is exactly what one would expect a priori, since lower-quality agricultural land would be cheaper and would often have greater scenic amenities for rural residential use.

These findings do not indicate that public policy should not be concerned with the rate of agricultural land conversion, but they do indicate that there is little basis in fact for the cataclysmic rhetoric which has sometimes characterized recent discussions of the agricultural land issue.

6

Provincial rural and agricultural land use policies

During the 1960s provincial policies concerning rural land use were aimed primarily at restricting scattered rural non-farm residential development and ribbon development along highways because of the alleged adverse effects of such development patterns on the cost of public services, on highway performance, and on the environment. During the 1970s, however, the province has also developed policies with the stated aim of preserving good agricultural land for agricultural use.

In this chapter, following a description of provincial policies to control non-farm residential development in rural areas, particularly policies toward rural severances and estates, and of provincial policies concerned with preserving prime and unique agricultural land, we shall consider the most important case study of the province's policies toward agricultural land, namely the Niagara urban boundaries issue, and provide an economic analysis of these provincial rural and agricultural land use policies.

NON-FARM RESIDENTIAL DEVELOPMENT IN RURAL AREAS

Before 1946 there were almost no local or provincial restrictions on division of land or residential development in rural areas of Ontario, and very little division of rural properties occurred except adjacent to built-up urban areas. The main pattern of scattered rural non-farm residential land use evidently involved the purchase by wealthy people of entire farms as weekend residences.[1]

During the period between roughly 1948 and 1960, there was wide-spread scattered rural non-farm residential development on lots of one-quarter to two acres by people with more modest incomes. This led to the introduction and

1 Punter (1974), pp. 123, 129–30, 206.

progressive tightening of local and provincial controls over the creation of rural lots by severances. Although some municipalities began to impose controls in the early 1950s, the province began to intervene on a significant scale only in 1966. Controls operated to restrict the number and increase the minimum size of lots created, and consequently they appear to have had several important predictable effects on scattered rural non-farm residential development: (1) the rate of scattered rural non-farm residential development was restricted; (2) the prices of scattered rural non-farm residential lots and houses increased; and (3) people with modest incomes were disproportionately excluded from ownership of such properties.[2]

Evolution of provincial policy on rural severances and estates

The Planning Act of 1946 gave the provincial Community Planning Branch an important role in controlling the creation of residential lots by assigning it responsibility for approving registered plans of subdivision in areas under subdivision control. However, as the Act evolved during its first decade there were a number of provisions which limited provincial control over the creation of residential lots:

1. Provincial approval of subdivisions was required only in areas placed under subdivision control by a local by-law.
2. Lots in approved plans of subdivision could be divided without further provincial or local approval unless the plan was placed under part-lot control by a local by-law. In areas placed under part-lot control, lots could still be divided upon consent of the local planning board (or later committee of adjustment or land division committee) appointed by the municipal council.
3. Even in areas under subdivision control, where a local planning board (or committee of adjustment or land division committee) existed, there was no provincial control over severances which were not parts of plans of subdivision. Such 'consents,' usually involving creation of individual lots, were under local control.
4. Even in areas under subdivision control, creation of lots of 10 or more acres was exempt from provincial control. Creation of such lots was also exempt from local control unless the municipality passed a minimum-lot-size zoning by-law requiring a larger acreage.

A study of exurban development in Ontario reports that 'both the consent procedures and the ten acre lot exceptions were of immense importance to the development of exurbia, for they became the floodgates through which the demand for building lots in rural areas was filled.'[3]

2 Ibid., Chapter 4.
3 Ibid., p. 133.

Beginning in the mid-1960s, there was a shift toward greater provincial supervision of the creation of lots through increased control of both subdivisions and consents. The Planning Act was amended several times to give the Community Planning Branch, then located in the Department of Municipal Affairs, greater power.

In the mid-1960s, all consents in areas under subdivision control became subject to review and appeal to the OMB by the Department of Municipal Affairs.

In 1966 the Minister of Municipal Affairs announced the province's 'Urban Development in Rural Areas' (UDIRA) policy, which involved restraint of permanent residential development in rural areas and restriction of urban development to municipalities which were prepared to provide complete urban services. The UDIRA policy statement contained inter alia the following points:

As they relate to year-round urban development, these policies may be stated generally as follows:
1. Year-round, urban residential development should take place in municipalities that have adequate administrative organization to cope with urban problems; that are equipped for and are otherwise capable of providing and maintaining necessary urban services, including piped water, sanitary and storm sewerage, street maintenance, schools, and recreational facilities; and that have demonstrated a willingness to provide these services; ...

There are certain important, but limited, exceptions to this general policy. These are:
1. Estate development at low densities, where provided for in an official plan and zoning by-law.
2. A limited amount of filling-in in existing development that might not conform with the general policy, particularly in hamlets and other small settlements and on the periphery of urban communities, provided that the municipality recognizes and assumes its responsibilities for such development.[4]

According to the 1966 *Annual Report* of the Department of Municipal Affairs: 'An examination of the 871 subdivision plans submitted for approval in 1966 indicates a large number of proposals at variance with the stated [UDIRA] policy. A substantial portion of the 92 plans not recommended for approval were turned down due to inconsistency with the policy.'[5] The department increased the frequency of its appeals to the OMB (virtually all of which were successful) against locally approved consents for non-farm residences in rural

4 Ontario, Department of Municipal Affairs, *1966 Annual Report*, p. 21.
5 Ibid., p. 22.

areas which 'conflicted with government policy expressed in the "Urban Development in Rural Areas" statements by the Minister of Municipal Affairs.'[6]

The exemption from controls which applied to creation of lots of 10 or more acres was ended in 1968, and thereafter local consents which could be appealed by the Minister of Municipal Affairs were required for all severances in areas under subdivision control.

Applications for provincial approval of local official plans, implementing zoning by-laws, and registered plans of subdivision were required to meet an increasing number of standards imposed by the province. For example, in their official plans rural townships were required to include policies developed by the Community Planning Branch for severances and for plans of subdivision for 'rural estate residential development,' i.e., low-density, rural non-farm residences not serviced by municipal sewers and water supply.[7] The Community Planning Branch forced certain rural townships to place restrictions on the rate at which lots could be created by severances, and estate developments were typically required to have lots of two or more acres, to have homes with 1,500 or more square feet of floor space, and to be confined to areas with low agricultural capability which had substantial natural amenities.[8]

Beginning in the mid-1960s, rural townships which had not been placed under subdivision control by local by-laws were increasingly placed under subdivision control by order of the Minister of Municipal Affairs,[9] and in revising the Planning Act in 1970 the provincial government placed the entire province under subdivision control and placed all registered plans of subdivision under part-lot control.

In a number of rural townships where scattered non-farm development was occurring in conflict with the UDIRA policy, the minister imposed zoning restrictions to freeze land use pending adoption of land use controls by the townships. In 1972 the minister relieved a local committee of adjustment of its power to approve consents.[10] In 1974 committees of adjustment in municipalities without approved official plans lost the power to grant consents.

More recently, in 1975 the Ministry of Housing adopted a statement of planning guidelines on land severance for land division committees and

6 Ontario, Department of Municipal Affairs, *1969 Annual Report*.

7 James F. MacLaren, Ltd. (1975), p. 177.

8 Punter (1974), pp. 203, 184–5, 193–4.

9 The minister has the power to place areas under subdivision control or zoning restrictions. The power was usually used in areas without municipal councils, particularly where growth was anticipated as a result of resource development in northern Ontario, but on occasion the power was used in areas with municipal councils.

10 Ontario, Ontario Economic Council (1973), p. 16n.

committees of adjustment,[11] and the ministry applied pressure on local governments to adopt interim land severance policies by late 1975 in areas which did not yet have official plans. In 1977 the ministry issued guidelines for consents and for rural estate residential development.[12]

The following summary of current provincial policies on rural severances and on estate developments is based largely on these three sets of guidelines.

Current rural severance and estate development policies

(a) Rural severance policy

In 1975 the Ministry of Housing issued a revised severance policy.[13] In 1977 the Ministry of Agriculture and Food issued a green paper with additional details of the province's proposed guidelines toward severances,[14] and in 1978 these were adopted as government policy. The principal policies of the province are:

1. All new or amended local official plans must contain severance policies and policies concerning the preservation of agricultural land which are consistent with provincial policy. Municipalities without such official plans should adopt interim land severance policies.
2. Rural non-farm severances should not be allowed where they would lead to scattered isolated residential development or ribbon development along provincial highways.
3. Rural non-farm severances should not be allowed on prime or unique agricultural land, or where they would violate the Agricultural Code of Practice's recommended distances between non-farm residences and intensive livestock operations,[15] or where they might limit the expansion of viable farming operations.
4. Rural non-farm severances should not be allowed where they would lead to residential development on or near mineral or aggregate deposits.
5. Rural non-farm severances should be restricted to areas in or adjacent to existing hamlets or villages which are not located on prime or unique agricultural land.
6. In agricultural areas, the only permissible severances are those related to agricultural needs, such as where a lot is required for full-time farm help or a retirement home for a long-time farmer, or where farm consolidation has rendered one house surplus.

11 Ontario, Ministry of Housing (1976).
12 Ontario, Ministry of Housing (1977c, d).
13 Ontario, Ministry of Housing (1976). See also Ontario, Ministry of Housing (1977c).
14 Ontario, Ministry of Agriculture and Food (1977a).
15 Ontario, Ministry of Agriculture and Food (1976b).

7. Otherwise, a farm may be severed only when each resulting parcel will be a viable farm.
8. Severances as well as plans of subdivision should not be approved if they would lead to development on floodplains or lands subject to undue airport noise, or on small lots with septic tanks.

In addition to these major land use policies regarding severances, there are a number of policies of a more technical nature, such as that development should not take place by severances when a plan of subdivision would be more appropriate, or that the site must be suitable for residential use considering matters such as drainage and sewage.

(b) Rural estate residential development policy

As we have seen, since the 1966 UDIRA announcement, the province has had a policy, in principle at least, of restricting rural non-farm residential development to areas in or adjacent to existing hamlets or villages. Exceptions to this policy occur in the case of (1) estate residential developments and (2) infilling between existing residences. Estate residential developments are low-density developments which are not located in or very near existing urban centres and which are not serviced by municipal sewers and water supply. The Ministry of Housing has recently issued guidelines which update and publicize provincial policies which have evolved for estate developments over the past decade.[16] The basic policies are:
1. Estate developments are permitted only when provided for in an approved municipal official plan, and they should normally be carried out by plan of subdivision rather than severances.
2. Estate developments should not be located on prime or unique agricultural land or in other areas which would restrict agriculture, on or near aggregate or mineral deposits, or anywhere that they would have a significant detrimental impact on the environment.
3. Estate developments should be located in areas with natural amenities and a rural character, and they should be located and designed so that they do not interfere with the rural landscape.
4. Estate developments should not be large enough to require their own schools, commercial outlets, etc.
5. No estate lot should be less than one acre, and the average density of an estate development should not exceed two persons per acre.

Provincial control of consents in practice

In spite of all the policy developments outlined above, it is evident that land severance activity continued to permit rural non-farm residential development

16 Ontario, Ministry of Housing (1977d).

which conflicted with the stated provincial UDIRA policies.[17] The following paragraphs will cite evidence that implementation of the UDIRA policies has been very incomplete, that is, that there has been a wide gap between provincial pronouncements and actions. However, this does not mean the provincial policies had no effect. Indeed, casual evidence suggests that rural land use controls did have a significant effect on rural non-farm residential development. The evidence is of two types. First, the price per acre of small unserviced rural lots is substantially higher than the price per acre of complete farms in the same area. Second, Russwurm's data summarized in Chapter 5 indicate that the absolute as well as the percentage increase in scattered rural non-farm population was lower in 1961-71 than in 1951-61, a fact which Russwurm attributes at least partially to planning controls.

We now turn to the evidence that implementation of the UDIRA policies was incomplete. A 1975 review of the UDIRA policies reported: 'Whether the policies were being implemented by the Province, the municipalities or their agencies (in particular, land division committees and committees of adjustment), they were seldom applied consistently, being more often used indiscriminately as convenience warranted, and otherwise disregarded or ignored.'[18]

As one illustration, a study of severances granted by the land division committee and by committees of adjustment in Lambton County in 1971-2 reported that the number of developable parcels created by consent was equal to the number created by registered plans of subdivision.[19] The study also states that '94.1% of the applications in 1971 to create new residential parcels in agricultural areas were approved and 100% of such applications were approved in 1972 by the Lambton County Land Division Committee. Thus, a significant proportion of the new residential building lots created by severance, have been created in agricultural areas (39% in 1971, over 55% in 1972) ... The fact that between 60 and 70 percent of the new residential lots in agricultural areas were created for use by individuals *not* directly involved in agricultural activities is somewhat disturbing. The practice also contradicts the Provincial government's policies respecting urban development in rural areas' (p. 21). The study goes on to report that 'many committees are, at present, not regulating adequately the size of severed and retained farm parcels, so that questionably small units (25

17 It has also been claimed that the creation of lots through severances has gone on virtually unabated in the Niagara Escarpment Area while the plan for the escarpment area is being developed by the Niagara Escarpment Commission. See Bureau of Municipal Research (1977), p. 84n.

18 James F. MacLaren, Ltd. (1975).

19 County of Lambton, Planning Department (1973), p. 4.

acres or less) are being created ... Some committees have been very restrictive in permitting consents for residential dwelling in rural areas ... Other committees have adopted far more permissive policies in which residential development can occur almost anywhere in rural areas on very small parcels' (p. 48).

As another illustration, a planning study in the Regional Municipality of Niagara reports that the main type of non-farm development in rural areas is residential development on lots created by consents, that the vast majority of rural lots have been and continue to be created by consent rather than by registered plans of subdivision, and that 'the resultant pattern of development is randomly located urban type uses, much of which is in the form of ribbon development along rural roads.'[20]

One explanation for the fact that rural development has not been consistent with UDIRA policies is that the sheer volume of consent applications has prevented the ministry from exercising close control over consents, be they consents granted by local committees of adjustment and land division committees on the one hand, or consents granted directly by the minister in areas without local committees on the other.[21] For example, in 1966 a total of 3,097 applications for consents in areas without local committees were directed to the minister, and only 6 per cent were rejected. It was reported that: 'of the 3,097 applications, 188 were not recommended for approval. The largest number not approved were applications to create sites for permanent urban occupancy at some distance from existing urban settlements. This action is consistent with the Minister's [UDIRA] statement relating to provincial policy on the distribution of urban uses in rural areas.'[22] Also, it was reported that in the case of consents granted by local committees:

A 1968 study revealed that, if all desirable appeals were initiated, the [Community Planning] Branch would average 40 cases a week, each of which would require an OMB hearing. Of necessity the Branch has concentrated its energies in the worst areas [i.e., in municipalities where more than 100 consents are granted per year, or where attempts are made to create whole plans of subdivision by the consent process] and in cases where the Minister is almost certain of winning.

In 1971 committees of adjustment granted over 12,700 consents. Of these, only about 150, or a little over one percent, were appealed by the Minister, chiefly

20 Regional Municipality of Niagara, Planning and Development Department (1977d), p. 1.

21 Generally, in southern Ontario consents are granted by local committees and in Northern Ontario consents are granted by the minister.

22 Ontario, Ministry of Municipal Affairs, *1966 Annual Report.*

TABLE 6.1

Consent activity in Ontario, 1973-6

Year	Ministerial consents		Local consents	
	Applications	Granted	Granted locally	Appealed by minister to OMB
1973	2,974	2,040	26,130	75
1974	1,488	916	21,566	42
1975	2,016	1,179	22,222	58
1976	1,899	1,150	17,472	15

SOURCE: Ontario, Ministry of Housing, *Annual Reports*, 1974-5 to 1976-7

because there is simply not enough staff to tackle anything but the very worst outrages. Besides, in the same year the minister granted a further 5,900 consents in municipalities without committees of adjustment or in unorganized territories where consents are administered directly by the Province through the Community Planning Branch. At the same time some 1700 applications were not approved.[23]

Table 6.1 gives recent data on the number of consents granted and ministerial appeals to the OMB in the case of severances granted by local committees.[24] The large number of consents granted and the small number of ministerial appeals to the OMB are striking.

PROVINCIAL POLICY TOWARD AGRICULTURAL LAND

For several years the provincial government has encouraged local governments to include in their official plans provisions for the protection of good agricultural land. In 1974 and 1975, the preservation of agricultural land became an important political issue because of publicity given to data on the rate at which land was being withdrawn from agricultural use between 1966 and 1971, escalation of food prices, and pressure on the provincial government by the opposition to increase provincial protection of agricultural land. A Food Land Development Branch was established in the Ministry of Agriculture and Food in 1974 to review official plans, subdivision plans, and other land use decisions and projects

23 Ontario, Ontario Economic Council (1973), pp. 16 and 60.

24 Not all consents involve creation of new residential lots; some severances involve minor readjustments in the boundaries of existing parcels, for example, and some involve non-residential parcels.

for conformity with provincial policy regarding agricultural land, and in 1975 guidelines on use of agricultural land were sent to the municipalities in the province. In 1976 and 1977 the Ministry of Agriculture and Food issued a general statement of provincial policy on agricultural land and a green paper outlining proposed guidelines for planning for agricultural land by local governments,[25] and in 1978 the guidelines were adopted as government policy. The policies announced in these two statements are summarized below.

The principal stated aim of provincial policy is to preserve the better agricultural land so that it will remain available for agriculture in the future, whether or not it is now used for agriculture. The province considers the better agricultural land to include:

1. All lands which have a high capability for the production of specialty crops because of the special soils or climate, such as the Niagara fruitlands, the Holland Marsh, and the Georgian Bay apple area.
2. All lands where Canada Land Inventory soil classes 1 to 4 predominate.
3. Additional areas where farms exhibit characteristics of ongoing viable agriculture.
4. Additional areas where local market conditions ensure agricultural viability where it might not exist otherwise, such as those adjacent to major urban markets and in northern Ontario.

The policy of preserving the better agricultural land involves several forms of restraint on division of land and on land use:

1. Non-farm use of land (other than forestry, open space, and certain types of recreation which do not conflict with agriculture and which permit the land to revert to food production) is to be restrained in areas of good agricultural land because:

(a) Land used for non-farm purposes is often permanently lost to agriculture.

(b) Non-farm use of land in agricultural areas restricts agricultural use of surrounding land because of conflicts between farm and non-farm land uses, such as complaints by non-farm residents about noise, smells, and pesticides from neighbouring farms, and the use by non-farm residents of local by-laws to restrict farm operations.

(c) Non-farm use of land in agricultural areas may prevent the expansion or consolidation of farm operations.

2. Farms are not to be divided into areas smaller than would be required for viable farms, so that future agricultural use will not be constrained by problems of reassembling land.

There are two primary land uses which the province is proposing to restrain: built-up urban use and non-farm rural use. In the case of built-up urban use, the

25 Ontario, Ministry of Agriculture and Food (1976a, 1977a).

province proposes to encourage three alternatives to development on good agricultural land:

1. As far as possible, urban growth should be restricted to areas within present urban boundaries.
2. Whenever possible, urban growth should be diverted to lands with low capability for agriculture in either the same or other communities. The province has stated that 'for municipalities which shift development from better lands to less favoured soils, within an urban boundary, special programs will be considered to compensate for higher servicing costs.'[26]
3. If necessary, urban development should take place at higher density.

In the case of non-farm rural uses, the province has already revised its policy on rural severances. As we explained above, it is now provincial policy that rural non-farm severances should not be allowed on prime (classes 1 to 4) or unique agricultural land, that rural non-farm severances should be restricted to areas in or adjacent to existing hamlets or villages which are not located on prime or unique agricultural land, and that in areas where land of marginal quality is limited in extent the amount of non-farm development will have to be curtailed.

In order to implement provincial policy on agricultural land, the province is relying primarily on the action of local governments to designate and protect agricultural land in their official plans and other land use decisions, subject to the powers of approval and appeal given to the province by the Planning act. Among other things, the province has indicated it will not approve local official plans and amendments unless they conform to provincial guidelines concerning agricultural land. The provincial government has been strongly criticized by the Liberal and New Democratic parties as well as groups such as the Bureau of Municipal Research[27] because of the province's decision to rely on guidelines for local government action rather than direct provincial legislation to protect agricultural land. For example, in 1977 the leader of the opposition called on the government to adopt a land use plan for the province with designation and legislative protection of good agricultural land, but this approach has been rejected by the government.[28]

On more specific provincial land use decisions, the government has also been criticized for failure to give adequate attention to preservation of agricultural land. The Toronto-Centred Region plan and the COLUC report[29] designated a number of second-tier communities for future growth on prime agricultural land. The Treasurer of Ontario supported the Barrie annexation proposal in 1976-7 in

26 Ontario, Ministry of Agriculture and Food (1976a), p. 13.
27 Bureau of Municipal Research (1977).
28 Ibid., pp. 31, 38–9.
29 Ontario, Central Ontario Lakeshore Urban Complex Task Force (1974).

spite of the fact that it included 7,000 acres of class 1 farmland.[30] Most (74%) of the land in the Townsend new town site is class 1 and 2 farmland.[31] The cabinet decided to allow an amusement part on good farmland at Maple.[32] In its 1977 Niagara urban boundaries decision, the provincial government accepted an urban designation for 4,000 acres of fruitland.

The Niagara urban boundaries issue, which is the most important test of provincial agricultural land policies to date, will now be examined in detail.

NIAGARA URBAN AREA BOUNDARIES ISSUE

In early 1977 the province decided that the area designated for urban development by the Regional Municipality of Niagara in its proposed official plan contained too much tender fruit, grape, and prime agricultural land, and the province cut back the urban area boundaries defining the land where urban development could take place. This case is the principal test of the province's new policy of preserving the best agricultural land for agriculture. We here recount the events leading up to the 1977 cutbacks and assess their impact on the preservation of the Niagara fruitlands.

Niagara fruit belt

The orchards and vineyards of the Niagara fruit belt are located along the south shore of Lake Ontario between Hamilton and the United States border, primarily in five municipalities which lie within the Regional Municipality of Niagara: Grimsby, Lincoln, Pelham, St Catharines, and Niagara-on-the-Lake (see Figures 6.1 and 6.2). The acreages of land under tree fruit and grape production in 1931-76 are shown in Table 6.2. According to the *Fruit Tree Census* in 1976, the Niagara Peninsula (the Regional Municipality of Niagara, the Regional Municipality of Hamilton-Wentworth south of Highway 99, and the former Haldimand County) accounted for 70 to 87 per cent of each of the principal types of tender fruit trees and 99 per cent of the grape vines in Ontario.[33] It will be seen below that the tender fruit and grape land at issue in the Niagara urban area boundaries discussion in the mid-1970s amounted to about 11 per cent of the maximum (1951) total area cultivated for tree fruit and grapes in the Niagara fruit belt.

The Niagara fruit belt has been undergoing extensive urbanization for several

30 Bureau of Municipal Research (1977), p. 34.
31 Ibid., p. 30.
32 *Globe and Mail*, 4 July 1978, p. 5.
33 Ontario, Ministry of Agriculture and Food (1977b).

TABLE 6.2

Niagara fruit belt acreages, 1931-1976

	Acres				Percentage change		
Fruit crops	1931	1951	1971	1976	1931-51	1951-71	1971-6
Peaches	7,200	14,100	9,300	n.a.	+96	−34	n.a.
Cherries	1,800	4,200	3,200	n.a.	+133	−24	n.a.
Apples	4,600	2,200	1,500	n.a.	−52	−32	n.a.
Pears	1,900	5,600	5,200	n.a.	+195	−7	n.a.
Prunes and plums	2,600	4,700	2,100	n.a.	+81	−55	n.a.
Total tree fruits	18,100	30,800	21,300	20,300	+70	−31	−5
Grapes	14,000	20,400	21,900	23,900	+40	+7	+9
Small fruits	1,300	2,000	300	n.a.	+54	−85	n.a.
All fruit crops	34,000	53,200	43,500	n.a.	+56	−18	n.a.

n.a.: not available

SOURCE: Krueger (1978b), Table 1, and Statistics Canada, Census of Canada, *Agriculture*

decades. The population of the five fruit belt municipalities of the Regional Municipality of Niagara increased from 64,200 in 1941 to 162,300 in 1971.[34] According to a study by Gierman, between 1966 and 1971 about 2,000 acres of former orchard and vineyard land was converted to built-up urban use in the Niagara region.[35]

It should be pointed out that the lands currently used to produce fruit in Ontario are not the only areas where fruit production is possible. The acreage used to produce fruit is determined by economic considerations, not simply by supply of land. Thus, grapes can be grown on a fairly wide variety of soils, and grape acreage and output in Ontario depend more on protection of the local wine industry by tariffs and the LCBO's discriminatory markups than on competition with urban development.[36] For example, grape acreage in the Niagara fruit belt has been increasing over the past several decades, and grape acreage has also been expanding recently in Essex and Kent counties in southwestern Ontario. Essex and Kent counties were important grape growing regions until the 1930s when most farmers switched to corn production.[37]

34 See also the summary of Krueger (1978b) in the preceding chapter.

35 Gierman (1977), Table 3.

36 See Acheson (1977).

37 *London Free Press*, 30 November 1978, p. B13.

Figure 6.1
The Regional Municipality of Niagara and the Niagara Fruit Belt

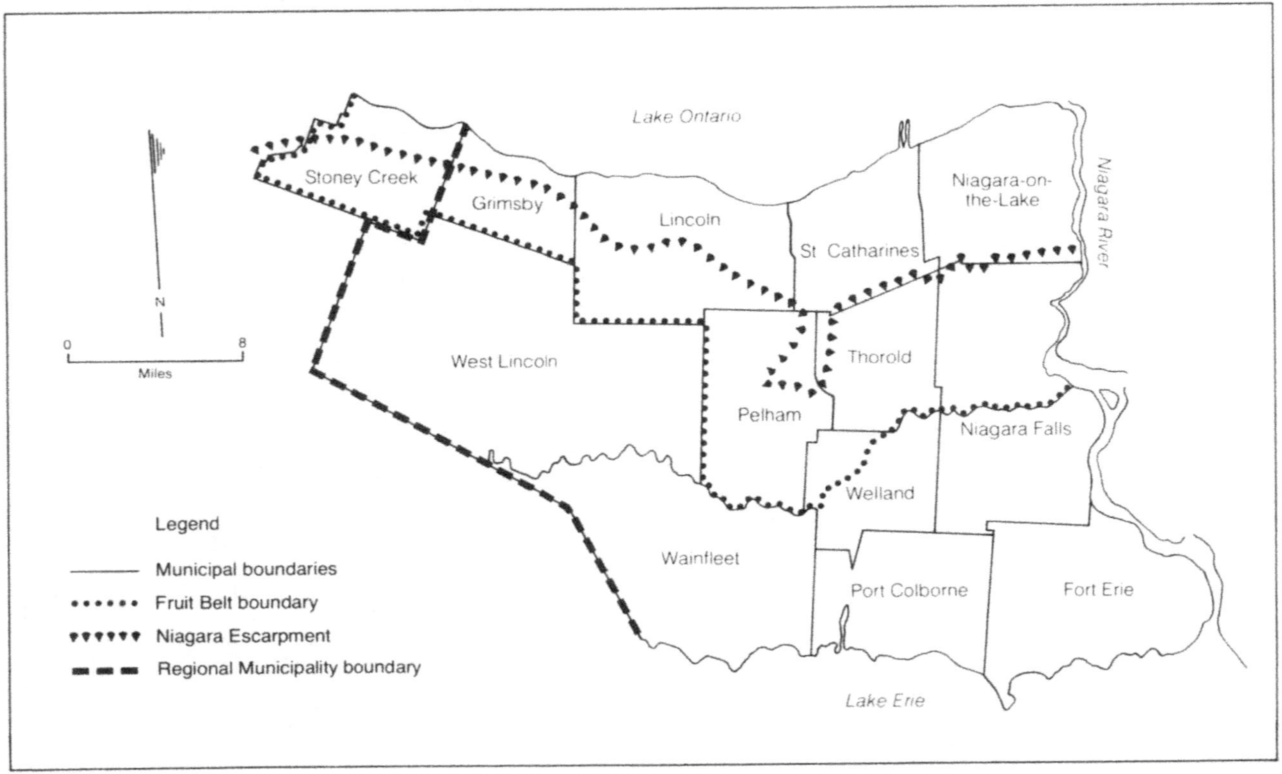

Figure 6.2
Orchards and vineyards as percentage of township block area, 1975
(from Krueger, 1978, Figure 4)

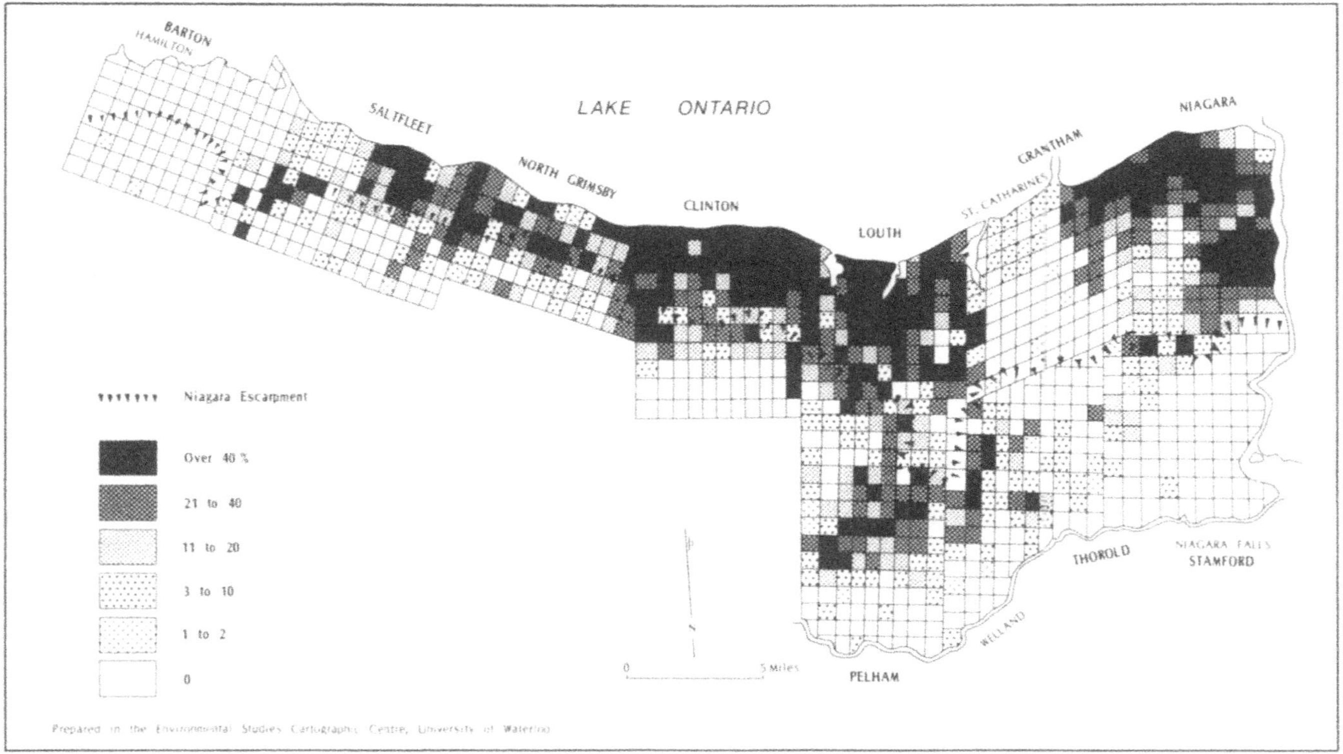

BARTON
HAMILTON
SALTFLEET
LAKE ONTARIO
NORTH GRIMSBY
CLINTON
LOUTH
GRANTHAM
ST. CATHARINES
NIAGARA
THOROLD
NIAGARA FALLS
STAMFORD
WELLAND
PELHAM
0
5 Miles
Niagara Escarpment
Over 40 %
21 to 40
11 to 20
3 to 10
1 to 2
0
Prepared in the Environmental Studies Cartographic Centre, University of Waterloo

TABLE 6.3

Land within 1974 urban area boundaries in the Niagara fruit belt[a]

Municipality	Total developable lands			Good tender fruit and grape land			Good general agricultural land[b]		
	Total	Residential	Ind. and comm.	Total	Residential	Ind. and comm.	Total	Residential	Ind. and comm.
St Catharines	3,595	2,325	1,270	2,855	1,865	990	–	–	–
Grimsby	1,190	795	395	855	585	270	95	85	10
Lincoln	1,440	1,185	255	1,220	965	255	160	160	–
Niagara-on-the-Lake	965	725	240	290	290	–	620	380	240
Pelham	860	860	–	710	710	–	150	150	–
Total	8,050	5,890	2,160	5,930	4,415	1,515	1,025	775	250

a These are the urban area boundaries as of March 1976, before the regional municipality produced the revised August 1976 boundaries in response to the September 1975 provincial objections. The March 1976 boundaries were very close to those originally proposed in October 1974.

b Class 1 and 2

SOURCE: Regional Municipality of Niagara, Planning and Development Department (1976), p. 13

Niagara urban area boundaries

In the second half of the 1960s two major provincially sponsored studies recommended regional planning for Niagara, with preservation of the fruitlands as one goal of a regional plan. In 1966 the Niagara Region Local Government Review (Mayo) Commission identified the preservation of the fruitlands from urban development as an important regional concern.[38] Two years later the Niagara Escarpment Study (Gertler) *Fruit Belt Report* recommended that a maximum amount of tender fruitlands be preserved for agriculture and that the province should prepare a regional plan.[39]

In 1970 the Regional Municipality of Niagara came into being as a result of provincial legislation, which required the region to submit an official plan by the end of 1973. The Regional Niagara Policy Plan submitted in 1973 included statements indicating that action would be taken to preserve the fruitlands, but until 1974 the region did not submit details of the urban area boundaries defining the limits of urban development.

The urban area boundaries proposed by the region in 1974 for the five fruit belt municipalities included about 8,050 acres of developable land, including 5,930 acres of good tender fruit and grape land and 1,025 acres of other good (class 1 and 2) general agricultural land (see Table 6.3). Jackson has argued that 'a substantial difference exists between (i) the Regional Niagara Policy Plan with its official statements about regional policy and (ii) the amount and location of land so designated for the purposes of urban expansion,'[40] that is, that although the regional plan included conservationist statements, supported by regional planners, the urban area boundaries proposed in 1974 reflected the expansionist goals of the municipal politicians.

In 1975, as a result of pressure from the New Democratic party, the urban area boundaries question became an issue in the provincial election, and the province refused to approve the urban area boundaries proposed by the region in 1974. A letter to the region from the Minister of Housing stated: 'The urban boundaries as presently shown in the proposed Regional Plan are unacceptable to the Province. I understand that about 5,200 acres of fruitland are designated for urban purposes in the Plan. This represents approximately ten percent of the total tender fruitlands in the province ... I am requesting Council to reconsider the boundaries of urban areas located on unique agricultural lands, in order to reduce substantially the encroachment of future urban development on this irreplaceable resource.'[41]

38 Ontario, Niagara Region Local Government Review Commission (1966).
39 Gertler (1968).
40 Jackson (1976), p. 68.
41 Cited in Jackson (1976), p. 73.

The region cut back on a number of the proposed urban area boundaries and submitted a new set of proposed boundaries in 1976. According to Jackson, these cutbacks involved deletion of about 500 acres of prime agricultural land from urban areas in the region.[42] In 1977 the province again refused to approve the proposed urban boundaries, and this time the province indicated the additional cutbacks it would require.

The province stated that the cutbacks it had imposed in 1977 amounted to about 3,000 acres. The leader of the New Democratic party claimed the cutbacks included only 1,780 acres, and according to Bacher the later figure is the correct one.[43]

The urban area boundaries specified by the province in 1977 were still subject to appeal to the Ontario Municipal Board, which began hearings on the Regional Niagara official plan in late 1978.

In announcing the 1977 cutbacks in the urban area boundaries, the Minister of Housing also stated that the province would offer two types of financial assistance to the region: (a) financial assistance for costs already incurred for servicing schemes which would not be used to full capacity because of the boundary reductions, and (b) financial assistance with additional future servicing and transportation costs incurred to redirect development from the better agricultural lands below the Niagara Escarpment to lower-quality agricultural lands above the escarpment.[44]

Effect of the boundary cutbacks on the fruitlands

Media reports on the urban area boundary cutbacks suggested that provincial intervention had saved something on the order of 3,000 acres of fruitland. However, reports prepared by the Regional Municipality of Niagara Planning and Development Department indicate that even the 1977 boundaries contain more than enough land to accommodate all the urban development that would have

42 Jackson (1976), p. 101. Bacher (1978) indicates that the 1976 regional cutback was 600 acres.

43 According to Bacher (1978), the provincial figure of 3,000 acres included some 600 acres which had been cut back not by the province in 1977 but by the region in 1976. It also included 620 acres of land which had already been developed and hazard lands which could not have been developed anyway. Bacher's calculations are consistent with a statement by the Regional Municipal of Niagara Planning and Development Department that the total reduction in the area within the urban area boundaries from the time they were proposed in 1974 through the cutback specified by the province in 1977 was about 3,000 acres. Evidently not more than about 2,300 acres of this was good agricultural land capable of future urban development.

44 Regional Municipality of Niagara, Planning and Development Department (1977b), p. 1.

TABLE 6.4

Niagara fruit belt, land capacity and housing requirements, 1976-96

Urban areas	(1) Vacant residential land[a] (acres)	(2) Dwelling units per acre	(3) Land capacity (dwelling units) (1) × (2)	(4) Expected requirement for dwelling units, 1976-96	(5) Population capacity 2.75 × (3)	(6) Expected population growth, 1976-96
St Catharines	1,720	8	13,760	n.a.	37,840	n.a.
Thorold	2,540	6	15,240	n.a.	41,910	n.a.
Subtotal	4,260		29,000	24,950	79,750	42,505
Grimsby	390	5	1,950	2,260	5,363	3,833
Lincoln	720	5	3,600	1,990	9,900	3,540
Niagara-on-the-Lake	700	5	3,500	1,740	9,625	2,515
Pelham	580	5	2,900	n.a.	7,975	n.a.
Welland	1,350	6	8,100	n.a.	22,275	n.a.
Subtotal	1,930		11,000	9,040	30,250	15,482
Total[b]	4,110	–	25,710	n.a.	70,703	n.a.

n.a.: not available

a 1977 urban boundaries

b For St Catharines, Grimsby, Lincoln, Niagara-on-the-Lake, Pelham

SOURCE: Regional Municipality of Niagara, Planning and Development Department (1977c)

occurred in the fruit belt municipalities during the next two decades even in the absence of boundary cutbacks.

According to Jackson's study of land use planning in the Niagara region, in 1974 and again in 1976 the region proposed urban area boundaries which included considerably more land, both in total and in virtually every urban area, than planners in the region expected would actually be used for urban growth during the next two decades, given even the highest projections of population growth and the lowest projections of population density for the region.[45] This proposal was made largely to satisfy the expansionist aims of the local councils in the region.

In announcing the boundary reductions in 1977, the Minister of Housing stated that 'provision can be made for urban growth below the Escarpment for

45 Jackson (1976).

more than 15 years on less land than proposed in your Official Plan. On this basis, we are cutting back the boundaries to more realistically reflect urban land requirements for the short term, say 10 to 15 years.'[46]

Subsequent reports by the Regional Municipality of Niagara Planning and Development Department indicate that, with the exception of Grimsby, the 1977 boundaries actually included more than enough land for development for at least 20 years (see Table 6.4).

The implication of this situation is that any savings of fruitland which will result from provincial intervention could be at least two decades away, and hence will be contingent upon maintaining the 1977 boundaries intact for a long time. Of course, if people believe that the 1977 boundaries will not be relaxed as the vacant land is developed, then even in the short run the price mechanism may lead to denser development within the urban area boundaries than would otherwise have occurred, i.e., the restriction on the supply of developable land could lead to higher expected future urban land rents and hence higher land prices and higher density of development. However, it is questionable whether people will expect the boundaries to remain intact indefinitely. For example, there is evidence in the case of the Niagara Escarpment that governments are unwilling to turn away developers who promise jobs in order to achieve government land use objectives.

ECONOMIC ANALYSIS OF PROVINCIAL RURAL LAND USE POLICIES

We have seen at the beginning of the chapter that since 1966 the province has had a policy of restricting rural non-farm residential development, and reviewed provincial policies to preserve agricultural land, including the Niagara fruitlands. A wide array of arguments have been presented, chiefly in the agricultural economics and planning literature, for government restrictions on rural non-farm residential development and the conversion of agricultural land to other uses. We shall now explain and evaluate the principal economic arguments for such restrictions, following the approach suggested in our summary of the methodology of economics in Chapter 2. We first analyse the arguments for restricting the conversion of agricultural land to other uses under the simplifying assumption that land is homogeneous, and then drop this assumption and deal with the arguments for restricting the conversion of agricultural land when land is heterogeneous. Finally, we analyse the arguments for restrictions on rural non-farm residential development.

46 Regional Municipality of Niagara, Planning and Development Department (1977b), Appendix I.

The economics of agricultural land conversion: homogeneous land
One of the issues in land use policy which has received the most attention in recent years is the social desirability of the existing rate of conversion of agricultural land to other uses. In Chapter 5 we reviewed the data on the rate at which agricultural land has been converted to urban and rural non-farm residential use and found that the rate is much lower than much of the public rhetoric has indicated.

(a) A simple model of the allocation of land
We shall begin with a simple long-run equilibrium model in which the allocation of land between agricultural and urban uses is determined.

The demand for agricultural land is a *derived demand*, i.e., it is derived from the demand for agricultural goods. Consequently, the determinants of the demand for agricultural land are the factors determining the demand for agricultural goods and the costs of agricultural production. Thus, an increase in the demand for agricultural goods, brought about perhaps by an increase in population, would lead to an increase in the demand for agricultural land. Similarly, the demand for urban land is derived from the demand for housing as well as the demands for the numerous other goods and services typically produced in urban areas.

In Figure 6.3 we illustrate the demands for agricultural and urban land, and the determination of the equilibrium allocation and rent of land. The base of the diagram has a width of $\bar{L}$ acres, the total amount of land. The demand for agricultural land, denoted *dd*, is drawn on axes with origin O. The demand for urban land, denoted *DD*, is drawn on axes with origin $\bar{L}$, and is measured *backwards*, i.e., from right to left. Equilibrium in the land market is at point e, the intersection of *dd* and *DD*, with equilibrium rent r^e, OL^e acres of land in agricultural use, and $L^e\bar{L}$ acres of land in urban use. It can be seen from the diagram that anything that would increase the demand for urban land, leaving the demand for agricultural land unchanged, would increase the rent of land, increase the urban use of land, and reduce the agricultural use of land.

Now let us consider the effect of land use controls. Suppose the government decides that at least OL^a acres of land must remain in agricultural use, where OL^a is greater than OL^e. It can be seen in the diagram that this control results in a new, lower equilibrium rent of agricultural land of r^d, and a new, higher equilibrium rent of urban land of r^{μ}. (We shall see later that when heterogeneous land is introduced into the analysis this rent differential may have an important impact on the efficiency of conversion of land of different agricultural productivities.)

This analysis, although simplistic, reveals two important points about the land market. First, anything that affects the demand for or cost of agricultural

Figure 6.3
Simple model of land allocation

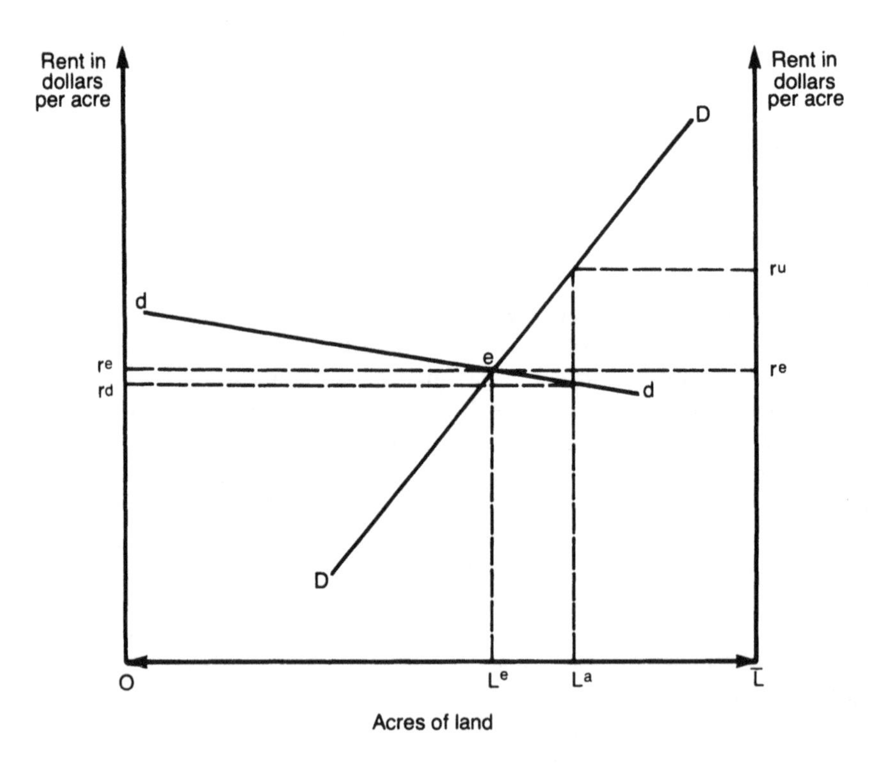

production or housing will affect the allocation and rent of land. Thus, for example, government programs having nothing directly to do with land use policies may have very important effects on land use and rents. One obvious example of such a policy is the protection of domestic agricultural production from foreign competition. This policy increases the demand for domestic agricultural production, and so, *ceteris paribus*, increases the demand for agricultural land. Similarly, construction and subsidization of commuter transportation systems, such as GO transit in the Toronto area, may increase the demand for land for urban use.

The second point revealed by our analysis is that land use policies affect not only land use but also land rents and hence prices. This obvious point is sometimes lost in public discussion of land use policies, where effects on land prices (and therefore landowners' wealth) are often ignored. We shall argue in the next section, for example, that many land use policies distort prices in a manner which may encourage the inefficient conversion of prime and unique agricultural land.

We must now consider the *rate* of conversion of agricultural land. A thorough analysis of the timing of land development, i.e., the rate of agricultural land conversion including the effects of market power, taxes, and uncertainty, is provided in Markusen and Scheffman (1977a, 1978b), and hence will not be repeated here. In a competitive land market without imperfections or government regulations, land conversion will proceed so that the rents of (raw) land in agricultural and urban use are equalized at every point in time. In a simple world with no uncertainty and perfect foresight, the price of land in either use will be equal to the present discounted value of (explicit or implicit) rents. Thus the price of land in agricultural use, for example, depends on the determinants of present and future demand for and costs of agricultural production and interest rates.

(b) Potential sources of market failure

As we explained in Chapter 2, in the absence of imperfections perfect competition will attain an efficient allocation of resources, i.e., resources will be allocated in such a way that no one can be made better off without someone being made worse off. In such an idealized world, the allocation of land between agricultural and other uses, and the rate of conversion of agricultural land to other uses, will be efficient. Other things being equal, a higher future demand for food will lead to a higher price for land and a lower rate of conversion of land to non-agricultural uses.

The possible economic case on efficiency grounds for government interference with the market rate of land conversion lies in the deviation of the real world from the idealized world we have been discussing. In Chapter 2 we explained that such deviations may result in a 'market failure,' i.e., the inability of the market to allocate resources efficiently. In this case an appropriate government policy may increase the efficiency of resource allocation. We shall now evaluate the various arguments which have been suggested for government restriction of agricultural land conversion based on market failures.

1. *Uncertainty* Even in the absence of other imperfections, the existence of uncertainty about the course of future events may result in an inefficient market allocation of resources. As we explained in Chapter 2, this inefficient allocation may occur if there are insufficient markets or institutional arrangements for risk-pooling. In the context of the agricultural land conversion issue, problems of predicting the future scarcity of agricultural products and the future urban

population are the main sources of uncertainty. Such uncertainty arises from the inability to predict with certainty the future path of population, technological change in agriculture, weather patterns, etc. An additional source of concern in this context is that it is expensive to convert land back to agricultural use. Thus, the cost of a 'mistake' of excessive conversion is high. However, it is *not* true, as many commentators implicitly or explicitly assume, that land converted to urban use is forever, under all conceivable circumstances, 'lost.'

Although very long term future markets for agricultural production do not exist, as always markets in real and financial assets offer the opportunity for considerable risk-pooling. One obvious method for someone to insure himself against future food scarcity is to buy farmland. In recent years it has become increasingly common for investment companies to include farmland in their asset portfolios. Thus it is possible for individuals to insure themselves (and their descendants) at least partially against future scarcity. Other things being equal, the demand for such insurance increases the demand for agricultural use of land, thus slowing the rate of conversion.

Society can purchase additional insurance against future scarcity of agricultural production by using government land use controls to restrict the rate of land conversion. However, consideration of such a policy should recognize that there are alternative forms of insurance, and that restricting land conversion is not costless. At least two major forms of insurance exist, and each is to some extent used. First, society can invest additional resources in agricultural research. Second, resources can be devoted to reducing population growth. Any policy chosen should recognize the alternative options and their costs. Unfortunately, restricting the rate of land conversion may seem less costly than other policies, because it requires no direct government expenditure (unless farmers are compensated). However, real costs are borne by society in the form of higher housing costs and lower housing consumption. Furthermore, the distribution of these costs and benefits is not uniform, since homeowners and some farmers (those who are able to sell their land for development) gain, and people who do not own homes and other farmers lose.

In summary, it is not clear that inefficient risk-pooling is a serious source of market failure. For various reasons (such as long-run international political stability) society may wish to increase its protection against future agricultural scarcity. However, the costs and the distribution of costs and benefits of restricting the rate of land conversion should be explicitly recognized and compared with those of other viable policies.

2. *Externalities* As we shall see below in our analysis of provincial restrictions on rural non-farm residential development, it has been argued that residential development in rural areas creates negative externalities which adversely

affect agricultural activities. However, this problem does not require a policy of restricting the aggregate rate of agricultural land conversion. If any government intervention is justified by externality arguments, it would take the form of local zoning-type controls on the location of non-farm development plus other policies such as local weed control by-laws.

3. *Agricultural land as a public good* It has been argued (Rodd and Van Vuuren 1975) that agricultural land is to some extent a public good, in that, apart from the benefit of using land for agriculture in terms of output of food, land in agriculture also benefits passers-by and others who enjoy the open space and the agricultural landscape. If this argument is valid the marginal social benefit of land in agriculture may exceed the marginal social benefit of land in housing at the market allocation, i.e., less than the efficient amount of land may be allocated to agriculture by market forces.

Four points should be considered in evaluating this argument. First, the empirical validity of the contention that agricultural land has value as open space has not been confirmed. Before any policy decision is made on the basis of the public good argument, there should be an attempt to assess the willingness of various segments of the populace to pay for the benefits and, as always, a careful consideration of the costs. Second, there may be positive externalities in the case of land use for housing as well; for example, an urban resident is probably better off if his neighbours have larger backyards, and hence people may tend to choose urban lot sizes below the efficient level. Third, it might be more efficient to subsidize agricultural land use rather than use direct restrictions on conversion of agricultural land to use for housing. Finally, we shall argue below that several existing federal and provincial policies already subsidize agricultural land use.

4. *Distortionary government policies* Various existing government policies have significant direct or indirect effects on the rate of land conversion. For example, the whole system of agricultural price supports, protection of agricultural goods from foreign protection, etc. raises the prices of agricultural goods and so slows the rate of agricultural land conversion, as does the favourable treatment of agricultural land by the property tax and succession duty systems. Other policies such as the construction of interurban transportation systems may increase the rate of conversion. Finally, it has been argued that unanticipated inflation and taxes on capital increase private discount rates, which may result in an increase in the rate of conversion.

(c) Equity concerns

In Chapter 2 we indicated that even when the market allocation is efficient it may not be optimal from society's point of view, because the allocation may not achieve society's concept of 'fairness.' In the context of land conversion policies, society's preferences relative to intergenerational equity are of prime importance. If we return to the simple world of the model which we considered earlier in this subsection, the market allocation (and rate of conversion) is efficient. The private discount rate (i.e., interest rates and financial market rates) is the result of the interaction of desires to save and to spend, and the pooling of risks. The level and direction of saving and the pooling of future risks will directly affect future generations, but these future generations do not have a direct 'vote' in either the market or the political system. Thus, society may wish to adopt policies which will increase the well-being of future generations. Such a situation is usually described as an excess of the private over the social discount rate.

It has been argued (e.g., Solow 1973) that a discounting of the future is not an ethically appropriate treatment of future generations, and that perhaps a social discount rate of zero per cent is appropriate. However, such arguments must be balanced against the effects of the many existing policies which affect the rate of agricultural land conversion (agricultural protection, etc.), the net effect of which is probably to slow the rate of conversion.

The economics of agricultural land conversion: heterogeneous land

In our previous discussion we assumed that all land is homogeneous, but much of the debate over preservation of agricultural land and much of the government's policy response have been built upon the heterogeneity of land. The basic concern has been with urban and other non-farm development on prime agricultural soils and on unique agricultural soils such as the Niagara fruitlands, and the government's major response has been to announce restrictions on non-farm development on prime and unique agricultural land.

(a) Preservation of prime and unique agricultural land

Many people seem to believe that all of the better agricultural land should be preserved for present and future agricultural use and that urban and other non-farm development should be restricted to inferior (e.g., classes 5 to 7) agricultural soils. In fact provincial land use guidelines have adopted this basic approach. There are two serious weaknesses in this naïve policy prescription. The first is that any rational policy toward preservation of agricultural land cannot possibly leave the aggregate quantity of land preserved to be determined by supply factors alone without regard to demand conditions; that is, there is no reason to believe that the total supply of classes 1 to 4 agricultural land and unique soils corresponds in any particular way to the socially optimal amount of land which should be preserved for agriculture. It is simply unacceptable to base provincial policy on slogans like 'You can never have too much good agricultural land.' Such reasoning totally ignores the opportunity cost of land preserved for agriculture.

The second weakness of the naïve policy prescription that the better agricultural land should be preserved for agriculture is that it ignores the fact that the opportunity cost (value in non-farm use) per acre of preserving different parcels of land for agriculture varies. There are two principal reasons for the variation. First, many of the physical characteristics which cause certain parcels of land to have a low capability for agricultural use, (because of problems related to slope, drainage, water table, etc.), or which raise the capital and operating costs per unit of agricultural output, also raise the same costs of using these parcels for urban use. Second, while location per se (at least within southern Ontario) is of little concern for most agricultural uses, it is a principal determinant of the value of land for non-farm use. For example, land located 5 miles from the centre of a large urban area sells for several times as much per acre for non-farm use as otherwise identical land located in a distant rural area. The price differential between otherwise identical land in different locations is largely a reflection of the capitalized value of the difference in transportation costs (e.g. commuting costs) for the parcels.

In dealing with the heterogeneity of land, proposed provincial land use guidelines have indicated that prime agricultural land should be preserved for agriculture and that urban development should generally be channelled onto areas with lower potential for agriculture. There is, in fact, no prima facie case for such guidelines, because they ignore the fact that the opportunity cost per acre of preserving different parcels of land for agriculture varies. Any decision to preserve a particular area for agriculture should be based on an explicit consideration of the costs as well as the benefits.

For example, consider a hypothetical example in which the government is considering preservation of 100 acres of class 2 agricultural land located 5 miles west of the centre of an urban area. In the absence of government intervention, the area would be developed for built-up urban use at a gross density of 5 households per acre. If the area is preserved for agriculture, suppose that the same new urban development would take place at the same density 10 miles south of the urban area on class 4 agricultural land which would have been used for agriculture in the absence of government intervention. (In reality, of course, the new urban development might be different in scale and density, the land in the existing urban area might be used more intensively, the population of the urban area might grow more slowly, etc.) In this case, the government restriction has three major effects: (1) the output of agricultural products will be greater and/or the total cost of labour, capital, fertilizer, etc., used in agricultural production will be lower because agricultural production takes place on class 2 rather than class 4 agricultural land; (2) the cost of production of housing services will (very likely) be higher, because of the higher cost of developing class

4 than class 2 land, due to problems of slope and drainage, and possibly because of the higher cost of providing sewer facilities south rather than west of the urban area, resulting from the longer distance to lake-based sewage treatment facilities; and (3) the cost (for roads, fuel, travel time, etc.) of commuting for a residential area 10 miles south of the urban area will be higher than the cost of commuting for a residential area 5 miles west of an urban area.

It should be obvious that the extra costs involved in locating a residential development 10 miles south instead of 5 miles west of an urban area are not trivial, and should be calculated and considered before government intervenes in a case like this. For example, the extra commuting costs alone in the example above would be quite considerable. Suppose (realistically) that from the 500 households 500 people would commute an extra 5 miles each way 250 times per year, with a car-bus modal split of 90–10, auto and bus occupancy rates of 1.2 and 20 respectively, auto and bus operating costs of $0.15 and $1.00 per mile respectively, an average speed of 30 miles per hour for both autos and buses, and an average value of travel time of $2.50 per hour for all commuters. The extra commuting costs would then be about $250,000 per year excluding the costs of the extra roads, or $2,500 *per acre per year.* To this must be added the costs of extra roads and higher development and servicing costs. It should be obvious that this would be a very high annual cost to pay for the extra output one would get from class 2 as opposed to class 4 agricultural land, with a given level of non-land inputs. (Indeed, the current market price of class 2 agricultural land in remote rural areas in southern Ontario is a good deal less than $2,500 per acre.) It seems likely that it would be more efficient to get the extra agricultural products by applying more input to other agricultural land, or by devoting more resources to research into improvements in agricultural technology, or by imports.

The preceding example is, of course, hypothetical, but it is not fanciful. It makes an extremely important point, namely that the cost of raising present and future agricultural output by preserving all the better agricultural land for agriculture may in some cases be high compared to the value of the extra agricultural output obtained or the costs of alternative means of obtaining additional agricultural products. If the government continues with the announced policy of preserving the better agricultural land for agriculture, it should at least provide explicitly for the policy to be waived in individual circumstances where it could be demonstrated that the costs would exceed the benefits.

It also follows from the preceding analysis that in an idealized world (perfect competition, no distortions) the price differential which good-quality agricultural land near urban areas commands compared to inferior land in less convenient locations reflects a savings in real resource costs in non-farm use. In such a world, as we have pointed out, the conversion of land will be efficient.

(b) Inefficient incentives for conversion of prime land

Government policies may create inefficient incentives for conversion of prime and unique agricultural land which would not be converted in an idealized world with no distortions. In particular, zoning and development controls themselves may create pressure for non-farm development of better agricultural land which would not exist in the absence of controls. In the simple model developed above we showed that zoning or development control restrictions create a price differential between land which is approved for development and land which is not approved, and this premium may make it profitable to convert the wrong parcels of land.

To see why government controls might create an incentive to convert the wrong parcels of land, suppose there are 50 acres of land which are homogeneous with respect to characteristics of developability (similar topography, location, etc.), but suppose that 25 of the acres are prime agricultural land and the other 25 are lower-quality agricultural land. Suppose that the values of the two types of land in agricultural use are $5,000 and $500 per acre, respectively, and that these values are independent of the development control policies of the urban area. Suppose further that in the absence of development controls the value of an acre of either type of land for development would be $3,000. In the absence of controls, the inferior acres would be converted but the good acres would not.

Now suppose that development controls are enacted which limit development to 10 acres. This policy will result in an increase in the value of land zoned for development. For example, if the value of an acre of land zoned for development increases to $6,000, the owners of each type of land will have an incentive to have their land zoned for development, although efficiency requires that only the inferior land be developed. Thus, where developable land of differing agricultural productivity is widely owned by small landowners, development control policies are likely to create some inefficiency in the rates of conversion of such land.[47]

In principle at least, this inefficiency could be corrected by appropriate government action. In our simple example, the best agricultural land could be zoned for agriculture. Two problems are immediately apparent in the possible enactment of a suitable policy. First, it is not clear that local governments would control

47 Similarly, the same development control policies may lead to inefficient fragmentation of rural land holdings, inefficient scattered, isolated rural residential development, etc., because the policies create an inefficient incentive to convert any and every parcel approved for development, regardless of the opportunity costs of the land in terms of value in agricultural use.

inefficient conversion. The property tax system would encourage the local government in our simple example not to allow the conversion of the better land, in order to maximize property tax collections, but the property tax rate and the importance of property taxes collected on agricultural acreage in urban areas are probably small enough to make such incentives weak relative to development pressures. Therefore, correction of such inefficiencies may require intervention by government agencies above the local level. The correct policy prescription in any particular case would, however, require a considerable amount of information. A simple policy of prohibiting the conversion of the best grades of agricultural land will *not* in general be appropriate.

Many economists and legal experts have advocated the setting up of some form of 'development rights' market as a means of enacting development controls or zoning. This sort of policy would seem to be especially suitable for the problem we have been discussing. It is beyond the scope of this study to go into the various development rights proposals in detail, but briefly such a system would work as follows. The urban government would determine how many acres of land will be approved for development and then would auction off the rights to develop. In our simple example, the owners of the less fertile land would bid more for the right to develop (since their profits from development would be higher), and so an efficient allocation (given the development controls) would result.[48]

Restrictions on rural non-farm residential development

A few of the arguments which have been presented for restrictions on rural non-farm residential development are fallacious on economic grounds, that is, they do not identify a potential source of market failure involving inefficient allocation of resources. Others do identify potential sources of market failure but, even if empirical analysis were to verify their importance, the corrective action called for would not involve a general provincial policy of restricting rural non-farm development but rather local policies such as spatial separation of land uses and weed control ordinances or provincial legislation to limit the power of local governments to impose restrictions on efficient agricultural activities.

Finally, some of the arguments (such as the argument that distortionary government policies subsidize rural development) identify potential sources of market failure for which provincial restrictions on rural non-farm development might be an appropriate policy response. However, the empirical importance of

48 However, if there are other significant distortions in the land market (such as road or sewer subsidies), such a system may not result in efficiency.

these alleged sources of market failure has not been established. Even if established, provincial intervention would still have to be justified by cost-benefit analysis. Moreover, provincial restrictions on rural development would be at best a 'second-best' response; the 'first-best' response would be to correct any distortionary policies which subsidize rural development.

(a) Cost differential for public services

The cost of providing public services to scattered rural residences is often alleged to be higher than the cost of providing the same services to an equal number of residences in a subdivision adjacent to an urban area. This claim is intuitively reasonable and is supported by a certain amount of evidence.[49] However, the empirical validity of the claim is beside the point, because minimization of the cost of providing a given level of public services is not a sensible criterion for the government to use in choosing between alternative residential location patterns. It might be efficient to have a significant level of scattered rural non-farm residential development despite higher costs of services if people were willing to pay the higher costs of services in order to live in rural areas. Arguments to restrict rural residential development in order to reduce the cost of public services should therefore be dismissed.

(b) Inefficient incentives for rural development

Another argument that has been used to support government restrictions on rural non-farm development is that there are other government policies which create inefficient incentives which encourage too many people to choose rural residences.

For example, it is sometimes alleged that the additional social cost of providing public services to people who live in rural areas is not borne entirely by these people, and hence rural residences are effectively subsidized relative to urban residences.[50] If this subsidization exists, the appropriate ('first-best') policy response would be a reform of public service user charges so that residences in various areas (urban subdivisions, isolated rural areas, etc.) pay the marginal social cost of their services.[51] If all types of residences paid this marginal cost, then people would choose rural residences only if they were willing to pay the extra servicing costs. However, in the absence of such a reform in user charges, it

49 Downing (1977).

50 Ibid.

51 The argument that user charges based on marginal social cost are 'first-best' assumes implicitly that the advantages of following such a policy in terms of allocative effects are not offset by disadvantages in implementation costs.

is possible that this argument would provide a justification for restrictions on rural development as a 'second-best' policy. Whether it would or not cannot be determined without careful study of the empirical validity of the argument and the benefits and costs of such restrictions.

Several other government policies have been alleged to create similar inefficient incentives for excessive rural development and hence have been cited in support of restrictions on rural non-farm residential development. For example, it has been argued that differences in local tax rates (perhaps supported by differences in provincial grants), and high urban land prices brought about by inefficient municipal zoning and development controls, provide inefficient incentives for people to live outside of cities. Controls over rural development have been suggested on the grounds that they would bring about a more efficient residential location pattern in the presence of these other policies, and on the more pragmatic grounds that they are necessary to support local taxation and land use policies. The comments we made above on inefficient public service user charges also apply to these other policies.

(c) Budgetary implications for local government

It is sometimes alleged that scattered rural non-farm residences impose a net budgetary cost on the rural municipality because the marginal cost to the municipality of providing services to scattered residences is greater than the marginal revenue raised by the rural municipality through user charges and taxes. Rural municipalities have responded to this budgetary situation (and other incentives) by imposing 'fiscal zoning,' i.e., large minimum lot and house sizes in order to make sure that property tax revenues for new non-farm residences will be sufficient to cover the costs of services provided to them and/or to maximize the budgetary surplus accruing to the municipality from new development.[52] Since rural municipalities have already adopted fiscal zoning, it is difficult to see how this argument could justify *provincial* intervention even if it were appropriate to control land use in order to improve the budgetary position of rural municipalities. Moreover, from a provincial point of view it is not sensible to choose among alternative land use patterns on the basis of their effects on local budgets. Choice among land use patterns should be based on considerations of efficiency of resource allocation. Where efficient land use patterns would involve budgetary problems for a local government, the appropriate response would be to remove the budgetary incentive to distort land use. This incentive might be removed by a change in the system of financing local services (through, for example, user charges, local taxes, provincial grants) or reorganization of local government.

52 Punter (1974).

(d) Political power in rural areas

As the proportion of non-farm residences in a rural area increases, the political power of non-farm residents increases relative to that of farmers. It is alleged that because of differences in tastes and incomes, the non-farm residents vote for higher levels of local services and taxes than farmers would choose, and consequently the farmers are made worse off. In addition, the non-farm residents may vote for restrictions on farm activities which cause noise, smells, dust, etc., which the non-farm residents find unpleasant.

Because non-farm rural residential development may have important implications for the distribution of well-being between farmers and non-farmers, many farmers (particularly those who do not have the opportunity to sell their land for urban development) support restrictions on non-farm development in rural areas. In assessing the distributional issue one should keep in mind that farmers already receive subsidies which reduce their local taxes, including lower assessments and rebates on property taxes.

It is also important to recognize the efficiency issues here. It must first be recognized that majority voting will not necessarily lead to the efficient service-tax package or efficient regulations on economic activity. Under simple assumptions, the public service level, etc. will be determined by the preferences of the median voter, and there is no particular reason to expect that this is the public service level which will maximize net social benefits for all the people affected. However, there is no obvious justification in terms of efficiency for the province to restrict land use in order to control the outcome of the voting mechanism. It would probably be more efficient for the province to restrict the power of municipalities to take the types of action which are found to be inefficient; for example, the province could limit local by-laws which restrict *efficient* farm operations. Of course, it should be kept in mind that some by-laws which restrict farm operations may increase the efficiency of resource allocation, and regardless of their efficiency they are likely to have important distributional effects. Consequently, the province should carry out a cost-benefit analysis before acting.

(e) Negative externalities

One problem of negative externalities between farms and non-farm residences is that farms impose a variety of negative externalities on nearby residences. One response has been local restrictions on farm operations. Another has been provincial specification, in the Agricultural Code of Practice, of minimum distances which must separate residences from intensive livestock operations. The Agricultural Code of Practice essentially allocates certain non-transferable property rights on a first-come-first-served basis, i.e., if a hog farm already exists in an area then new residences are prevented from locating nearby, and if a residence already exists in an area then a new hog farm is prevented from locating nearby. Although the restriction on hog farms locating near existing residences is a conventional method of dealing with negative externalities, it is a little more difficult to understand why the province should prevent a person from building a new residence near an existing hog farm if he wishes to do so. This regulation is, however, consistent with provincial restrictions on development on flood-plains and where airport noise levels are high. Presumably there are two reasons that the government does not accept the judgment of the market on which areas are suitable for development: (1) to protect unsuspecting buyers in a world of incomplete information, and (2) to prevent the development of pressure groups which would attempt to restrict hog farm and airport activity, demand compensation for flood losses, etc.

Non-farm residences may also impose various negative externalities on nearby farms. For example, non-farm residents may let weeds grow on their lots. This particular externality is typically dealt with by local weed control by-laws, however. Similarly, it is argued that some non-farm residents destroy farmers' fences and damage their crops, but such problems of vandalism are appropriately handled through the legal system.

It is also alleged that non-farm residences have a variety of less tangible adverse effects on the rural environment. For example, some people lament the changes in the rural way of life and the rural landscape when city people and non-farm residences appear in rural areas. Some observers also maintain that non-farm development which displaces agricultural activity leads to a decline in the size of the rural market for agricultural support services and farm labour pools and hence a decline in their availability, and consequently threatens the viability of agriculture.

The provincial government itself has often argued that non-farm residential development along provincial highways should be restricted because vehicles entering and leaving these residences impose negative externalities on other highway users, i.e., they slow down through traffic, sometimes to the point that the province builds a new highway to bypass the original development.

Non-farm rural residential development based on septic tanks may also increase pollution of ground-water resources and lakes. It should be kept in mind, of course, that agricultural activity also causes pollution; fertilizers, pesticides, and manure from livestock operations are, for example, significant sources of water and air pollution.

(f) Agricultural land

In addition to the various adverse effects which rural non-farm residential development is alleged to have on nearby agricultural activity, non-farm residential development may absorb farmland directly and may break up agricultural holdings and make it more difficult for existing farms to expand or for potential farmers to assemble large contiguous blocks of farmland in the future. We have discussed the economic rationale for restrictions on the conversion of agricultural land above. That analysis, which suggests that there is a *potential* case for actions to reduce the rate of conversion arising out of differences between social and private discount rates, applies to the present issue. It is important to recall, however, that most of the land used for rural non-farm residences typically is *not* prime (classes 1 to 3) agricultural land.

(g) Land prices

It has been alleged in the agricultural economics and rural planning literature that non-farm residences are incompatible with agricultural land use in the surrounding area because non-farm residential demand drives up land prices so high that agriculture is no longer viable. For example:

The sale of the single piece of land for, say, a new rural non-farm house will immediately create a shadow of higher values on a surrounding area much larger than the fragment actually used for the house. The opportunity cost for existing farmers will rise, and the real cash cost will be higher for new or expanding farmers ... The high purchase price for land will inhibit the entry of new farmers. It will inhibit the purchase of land to enlarge current farms. It will accelerate the exit of farmers who are less efficient, or older, or less emotionally committed to farming. The high price and the resultant uncertainty will inhibit investment in modernization by farm operators, gradually eroding competitive efficiency and productivity over wide areas.[53]

We have already analysed related issues in our discussion of 'Speculation and Rising Land Prices' in Chapter 5, but we consider the matter further here. In evaluating the argument in the preceding quotation, we shall consider the impact of non-farm demand on rural land rents and prices and rural land use in two situations: first, when there is no government restriction on rural non-farm development and, second, when the extent of such development is restricted.

We begin by assuming there is no government restriction on rural land use. Suppose there is a fixed amount of land which has two uses: agricultural and non-farm residential. Then the current rent on the land and the allocation of that land between uses will be determined by supply and demand. Normally, where the demand curves for land for both uses are downward sloping and each use succeeds in obtaining some of the land, one can conclude that the existence

53 Rodd (1976a), pp. 166–7.

of non-farm residential demand for rural land leads to a higher rent and a lower allocation of land to agriculture than would occur if there was no non-farm residential demand. However, the interesting empirical questions are *how much* land would be removed from agriculture and how much the rent would be raised in the absence of controls. Some of the agricultural economics and rural planning literature seems to suggest that there would be little land left for agriculture. Our own feeling is that such a suggestion is absurd. Even in the absence of planning controls the vast majority of rural land would be allocated to agriculture simply by the normal operation of supply and demand.

This case is made somewhat more complicated, but is not altered in substance, if we consider the future as well as the present. Suppose that as time goes by we expect that the non-farm residential demand curve for land will rise more than the agricultural demand curve. Then as time goes by the real rent on rural land will probably increase and the amount of land used for non-farm rural residences will increase. *Current* land rents or land use would not be affected, of course, but current land *prices*, which equal the present discounted values of future rents, would be affected.

Everything we have described here represents the normal, efficient functioning of a competitive market for a durable scarce resource. The effect of non-farm demand on rural land rents and prices is not, per se, evidence of a market failure. Moreover, the suggestion that non-farm residential development would lead to a situation where the price or rent on farmland was so high that virtually all rural land would either be used for non-farm residences or be left idle is simply untenable. As we saw above, there may be a potential case for government intervention to reduce the rate of conversion of agricultural land, but the case is not based on simple observation about the effect of non-farm development on rural land prices or rents.

Second, we must now consider the impact of non-farm demand for rural land when non-farm residential development is being restricted by government policy. The effect of such restrictions would be to raise the current rents (and future rents and prices) on land approved for non-farm residential use and depress the rents and prices on land not approved, i.e., that reserved for agriculture. The price of any parcel of agricultural land would depend, however, not only on agricultural rents but also on the probability of securing approval for conversion to non-farm residential use, on non-farm residential rents, and on the cost of securing the approval for land use change. In this case, the land use controls may create two problems, namely uncertainty in the land market and an inefficient incentive to try to secure approval for non-residential development on virtually all parcels of agricultural land. As we saw above, the implication of the latter problem is that planning controls may induce a waste of resources on the part of people seeking approval for land use change and may lead to a situation where non-farm residential development takes place on the *wrong parcels* of rural land. However, nothing in this argument suggests that rural non-farm residential development will lead to idling of rural parcels *not* approved for non-farm development.

In conclusion, we find no merit in the argument that non-farm rural development raises rural land prices and hence brings about the idling of rural land which is not actually used for non-farm development.

(h) Summary of arguments for restrictions on rural development

The preceding discussion of the arguments for government restrictions on rural non-farm residential development suggests that some of the common arguments are invalid but that others provide a *potential* justification for provincial land use controls. It is important to emphasize that no restrictions on rural development can be justified by this type of analysis alone. One must also conduct careful empirical studies to verify and quantify the alleged facts which underlie these arguments and carry out cost-benefit calculations to demonstrate that the benefits of such restrictions would outweigh the costs. Unfortunately, much of the rural planning literature ignores the importance of empirical verification and cost-benefit analysis. This literature often cites the hypotheses about the adverse effects of rural non-farm residential development and immediately proceeds to argue for tight provincial restrictions on such development, without the slightest concern for the costs of such controls.

(i) The nature of rural development restrictions

It should be emphasized that restrictions on rural non-farm residential development could take many forms, depending largely on the justification for the restrictions. For example, if the primary concern is the adverse effect of strip development along provincial highways on the speed of traffic, the province could limit strip development but permit rural subdivisions with a limited number of access points to provincial highways. If the primary concern is the loss of good agricultural land, the province could restrict rural non-farm development on prime agricultural land but permit it on inferior land. If the primary concern is the adverse budgetary effect of non-farm residences on rural municipal budgets, municipalities themselves could impose minimum-lot-size zoning. However, requirements for large lots could encourage withdrawal of land from agricultural use.

One of the more provocative suggestions regarding the nature of restrictions on rural non-farm development can be found in a report by James F. MacLaren (1975) and in a study by Rodd and Van Vuuren (1975). These studies argue that a viable agricultural system cannot exist side by side with non-farm residences. Consequently, they argue for spatial separation of farming and non-farm residences. Large blocks of the better agricultural land ranging in size from one square mile to an entire county would be reserved exclusively for agricultural activities, including activities which are supportive or neutral to agriculture; non-farm residences would be concentrated in hamlets and on soil with low capability for agriculture. Unfortunately, the studies which recommend this approach to rural planning have not been based on an explicit evaluation of the economic benefits and costs of the recommended actions, and hence it would be premature to base policy decisions on these recommendations. This approach, which would extend the traditional zoning mechanism to the rural countryside, does, however, deserve further study. The mechanism offers the potential of reducing farm/non-farm externalities without necessarily reducing the aggregate level of non-farm development although it might also be used to reduce this level if there was a justification for doing so. Of course, there would be costs as well as benefits in such a spatial separation of land uses; for instance, commuting costs would probably be increased.

CONCLUSIONS

We have seen that provincial rural and agricultural land use policies have two principal objectives: restriction of scattered non-farm residential development in rural areas and preservation of the best agricultural land for agricultural use. On the one hand it appears that provincial land use policies have had a significant effect on the allocation and prices of rural land. On the other hand, it appears that provincial policies have been implemented only partially, and hence there continue to be wide-spread land use changes which are in apparent conflict with stated provincial policy.

The quality of the economic analysis which has surrounded the formulation of provincial land use policy has been low. Upon inspection, one finds that some of the common arguments for provincial restrictions on rural land use simply would not justify provincial action to curtail rural non-farm residential development or agricultural land conversion. For example, the fact that rural non-farm residents may impose restrictions on efficient agricultural operations which they find unpleasant is an argument for provincial restrictions on the power of rural municipalities, not an argument for restrictions on non-farm land use. One also finds that existing government policies may create inefficient incentives for rural non-farm development and conversion of good agricultural land. For example, public service user charges should be set so that rural locations are not subsidized, and policies to restrict the aggregate rate of conversion of agricultural land should be accompanied by auctioning of development rights.

Throughout the literature on agricultural economics and rural planning one finds assertions that the allocation of agricultural land is too important to be left to market forces. For example:

Changes in rural land use cannot be viewed simply as a case of economic choice where we examine individual reactions to individual opportunities for gain and loss. Agricultural land must be examined from the social point of view, from the planning point of view, because the issues go beyond the dollars and cents of markets and the decisions by producers and investors relating to financial incentives and rewards.[54]

The price of land for non-farm related uses is traditionally higher in the free market place than is the price of land for agriculture. Consequently, agricultural land use planning has to be based on considerations other than price.[55]

One of our principal arguments in this monograph is that such wholesale recommendations to replace market allocations by planning allocations without careful consideration of the benefits and costs of government intervention should not be tolerated. There is no excuse for basing government policy on this sort of reasoning.

Such assertions have typically been followed by recommendations which involve replacement of allocation by the market by allocation by planners on the basis of ad hoc decision rules such as 'all the better agricultural land should be preserved for agriculture,' or 'all non-farm residential development should be excluded from agricultural areas.'

We have argued that although government intervention to influence the aggregate rate of agricultural land conversion and/or the spatial pattern of rural development *might* be justified in some cases by careful benefit-cost analysis or equity considerations, casual references to externalities in rural areas, public goods characteristics of agricultural land, or differences between private and social discount rates do not in themselves justify government intervention. Adoption of any general provincial land use guidelines or controls should be preceded by explicit consideration of benefits and costs and their distributions. Moreover, there should be provision for waiving any general restrictions in specific cases on the basis, among other things, of benefit-cost comparisons.

This is not to say that provincial land use policies should always be made on the basis of economic calculations alone. Decisions which have important effects

54 Ibid., p. 160.
55 James F. MacLaren, Ltd. (1975), p. 16.

on the allocation of scarce resources or distribution of well-being should not, however, be made without explicit consideration of costs and comparison of costs with benefits.

In 1978 the province went ahead and declared that it is official government policy to restrict agricultural land conversion by implementing the food land guidelines outlined in its 1977 *Green Paper on Planning for Agriculture.* Fortunately, the guidelines do at least provide that prime agricultural land may still be converted to other uses when the case for doing so can be justified. This provision is important, because it is fundamentally irrational to argue that all the better agricultural land should be preserved for agriculture, just as it would be irrational to argue that all the better hospital land should be preserved for hospitals or all the better airport land should be preserved for airports, without considering present and future demands for agricultural products, hospital services, or airport facilities and without considering alternative uses for the land in question for houses, schools, firms, etc.

7

Provincial regional planning and the Toronto-Centred Region Plan

Provincial involvement in comprehensive regional land use and development planning in Ontario became a matter of policy in 1966 with the announcement of the 'Design for Development' program.[1] The first product of the new provincial planning process was the plan for the Toronto-centred region (TCR), which was developed as a policy option by the Metropolitan Toronto and Region Transportation Study during the mid-1960s,[2] unveiled formally in 1970,[3] and adopted as policy in 1971.[4]

After the very general terms of the 1966 Design for Development regional policy statement, the announcement of the TCR plan in 1970 was important in the evolution of provincial land use policy not only because of the course of action to which the government appeared to commit itself for the TCR but because the government indicated that it intended to assume responsibility for a considerable sphere of economic activity which had previously been left to the private market and municipal governments.

We shall begin our discussion by reviewing the evolution of provincial regional policy up to the TCR plan, then outline the major features of the plan as they developed during the period 1970-8, and at the end of the chapter evaluate the province's regional planning experience.

As we shall see, although the TCR plan was declared to be provincial policy, the government never followed through on many of its features, and the province has abandoned much of the Design for Development approach to comprehensive regional planning. Thus, in many respects the status of the TCR plan as

1 Ontario, Government of Ontario (1966).

2 Ontario, Department of Municipal Affairs, Metropolitan Toronto and Region Transportation Study (1968).

3 Ontario, Government of Ontario (1970b).

4 Ontario, Government of Ontario (1971a).

'policy' was only nominal and of limited duration. To some extent this was probably recognized at the time by the politicians involved; indeed, one should normally discount government policy statements rather than accept them at face value because they contain material intended to secure media coverage and trial balloons which are not backed up by a genuine commitment of action.

Nevertheless, because the government went so far as to announce the TCR plan as policy, and because it was used to justify a number of subsequent actions, it is important to analyse the plan and to determine what lessons can be learned from this experience with provincial regional planning. In the following sections we examine the nature of government intervention in the economy which the province proposed and the economic and other problems which led to the virtual abandonment of the program. Obviously, any future consideration of regional planning in Ontario or elsewhere should benefit from the knowledge gained from the TCR experience.

BACKGROUND TO THE TCR PLAN

The province has long-standing policies to support the planning and provision of specific services such as transportation, recreation, and sewers at the local and regional levels. For example, in 1962 the provincial government established the Metropolitan Toronto and Region Transportation Study (MTARTS) to make recommendations concerning transportation planning for the Toronto-centred region. In 1967 the province sponsored the Niagara Escarpment Study to make recommendations concerning preservation of areas of the escarpment for recreational purposes in the face of increasing urban development and quarrying.

Some of these planning activities involved the province in regional land use planning, and a number of the most important features of the TCR plan developed from these earlier planning efforts. For example, the MTARTS 'Goals Plan II' suggested two rows of urban areas along the lakeshore, separated by a parkway belt,[5] and the Niagara Escarpment Study (Gertler) report recommended preservation of extensive portions of the escarpment for public recreational use.[6] In contrast, some of the provincial servicing programs during the 1960s encouraged a development pattern which the TCR plan sought to modify; for example, the Peel servicing schemes facilitated development west of Metro Toronto.

5 Ontario, Department of Municipal Affairs, Metropolitan Toronto and Region Transportation Study (1968).

6 Ontario, Department of Treasury and Economics, Regional Development Branch, Niagara Escarpment Study Group (1968).

Direct provincial involvement in comprehensive regional land use and development planning became a matter of policy in very general terms in 1966 with the White Paper *Design for Development*. This document asserts provincial responsibility for regional planning: 'The provincial government ... has the responsibility to carry out and give direction to regional land use and economic development planning ... The responsibility for the control and administration of any regional undertaking by the government should be in the hands of a central authority which can cut across both departmental lines and county or municipal boundaries in meeting and solving regional problems.[7] The province was divided into regions, for each of which a land use and development plan was to be prepared by the Regional Development Branch of the Ministry of Treasury and Intergovernmental Affairs (later TEIGA), which was given responsibility for the new regional planning program. The 1970 TCR plan was the first of these regional plans to be released.

The objectives and scope of the new provincial regional planning program were left vague, but evidently as of 1966 the policy was intended to involve a combination of (1) regional economic development policy to reduce economic disparities between the regions and (2) regional land use planning policy to bring about an 'orderly' spatial pattern of development within each individual region of the province.

LAND USE OBJECTIVES OF THE TCR PLAN

Basically the TCR plan is a regional land use plan for an area of about 8,600 sq. mi. (see Figures 7.1 and 7.2). The region is divided among various broadly defined land uses (urban areas, parkway belts, agriculture, recreation, conservation), and rough population targets for the year 2000 or the early part of the next century are assigned to the various subregions and, in some cases, urban areas.

Before proceeding with our outline of the major land use objectives of the provincial government's plan for the TCR,[8] it will be useful to introduce a few of the terms used in TCR planning to describe various subregions of the TCR (see Figures 7.1 and 7.2). First, the 'lakeshore urbanized area' (Zone 1) refers to the subregion along Lake Ontario from Hamilton to Oshawa. Second, the 'commutershed' (Zone 2) refers to the subregion immediately north of the lakeshore urbanized area and stretching roughly from Acton (Halton Hills) to Port Perry. Finally, the 'peripheral zone' (Zone 3) refers to a semicircular ring, including

7 Ontario, Government of Ontario (1966), pp. 3-4.
8 Ontario, Government of Ontario (1970b, 1971a).

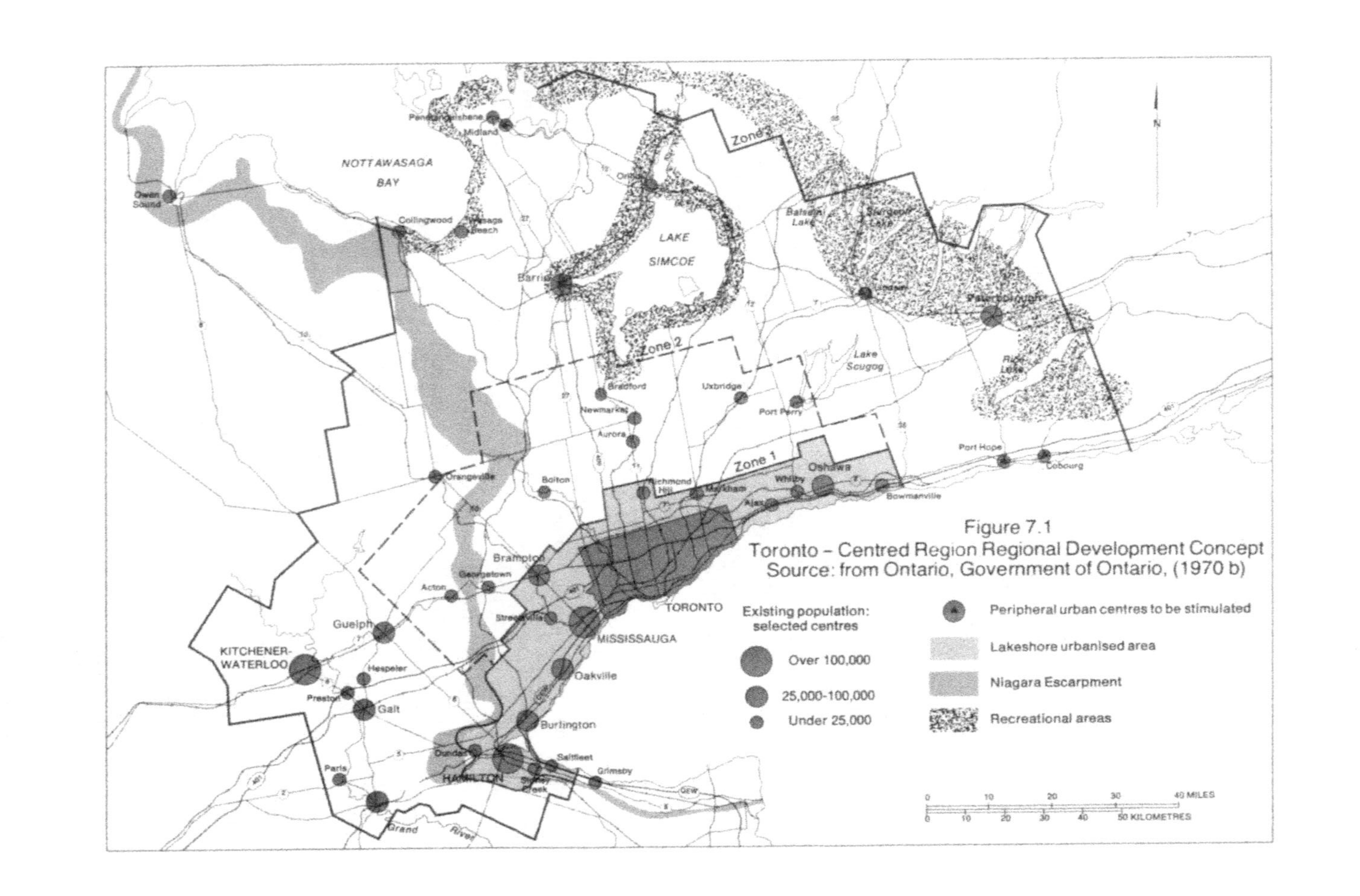

Figure 7.1
Toronto – Centred Region Regional Development Concept
Source: from Ontario, Government of Ontario, (1970 b)

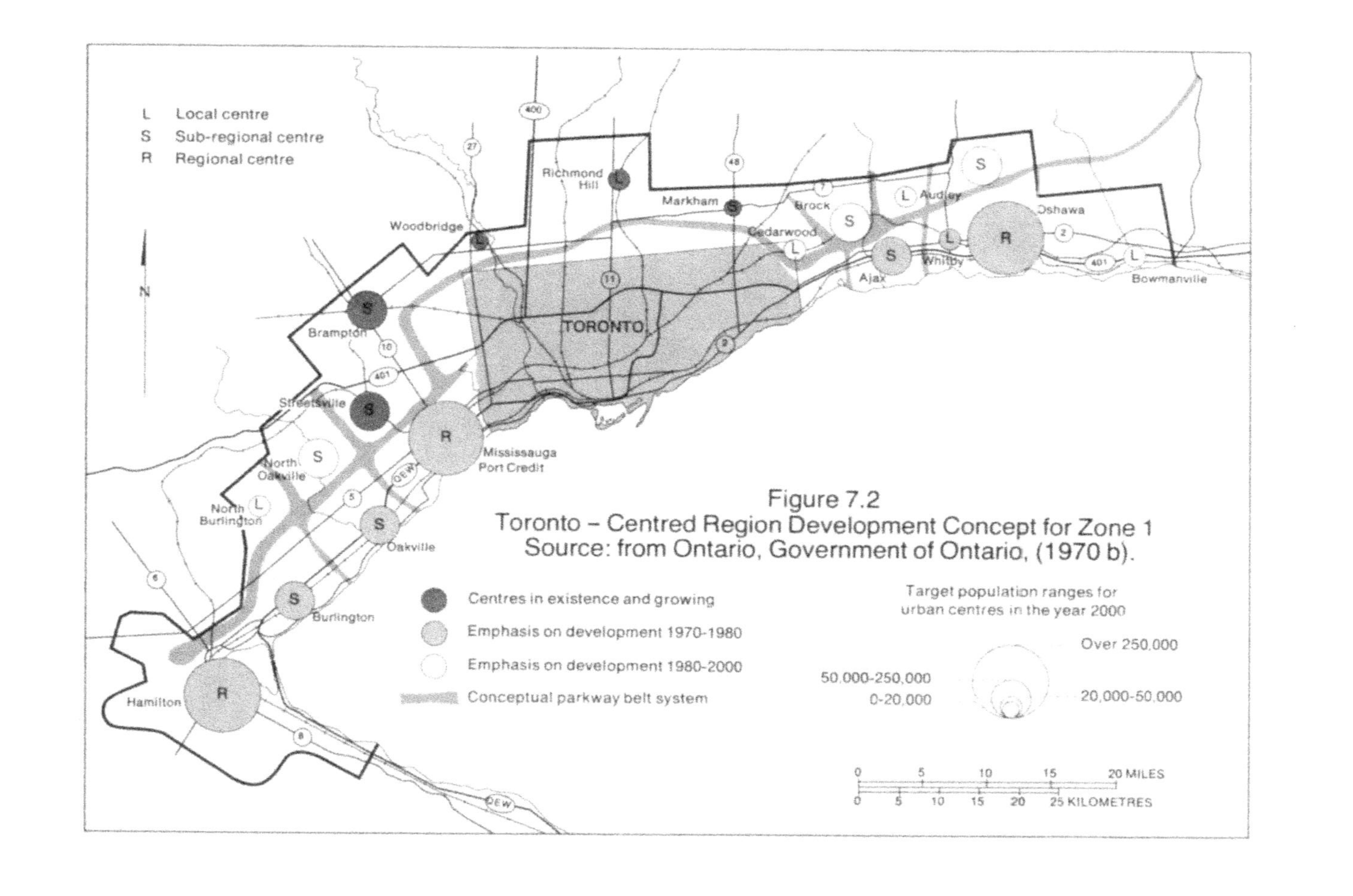

Figure 7.2
Toronto – Centred Region Development Concept for Zone 1
Source: from Ontario, Government of Ontario, (1970 b).

Kitchener-Waterloo, Barrie, and Peterborough, which is outside commuting range to Toronto.

Size and spatial distribution of urban areas

One of the major purposes of the TCR plan is to bring about a size and spatial distribution of urban areas in the region different from what would otherwise evolve. The basic features are:

1. *Decentralization and 'go-east'* Redirection of population and employment growth to urban areas in the eastern part of the lakeshore urbanized area and in the northern and eastern parts of the peripheral zone (Midland, Barrie, Port Hope, Cobourg) and the remainder of the province. Although there was no suggestion in the plan that the growth of Metropolitan Toronto should be restrained, it is generally agreed that the intention was to take some of the development pressure off the lakeshore area by designating growth centres in the north and east well beyond commuting range of the lakeshore area.

2. *Two-tier linear urban structure and parkway belt* Redistribution of growth within the lakeshore urbanized area into a linear (rather than radial) pattern, with two parallel rows of urban areas, one along the lakeshore and another inland a few miles to the north. A number of the inland areas would be new communities. The two rows would be separated by a parkway belt, and adjacent urban areas in each row would be separated by mini-belts into distinct, identifiable communities. The distributions of population and employment among the twenty or so urban areas involved would form deliberate size and functional hierarchies.

3. *Balanced growth* Encouragement of balanced growth of population, employment, and services in each urban area within the lakeshore urban complex, to avoid creation of dormitory suburbs.

Preservation of land for agriculture, recreation, and open space

The second major purpose of the TCR plan is to reserve large areas for purposes other than urban, residential, cottage, hobby farm, and 'speculative' uses. It was expected that implementation of this policy would involve significant restraints on urban and residential development in some areas. The major areas to be reserved are:

1. The commutershed (Zone 2).
2. The parkway belt system in the lakeshore urbanized area, to be reserved as open space with mainly non-urban uses but to contain interurban transportation and utility services.
3. Recreation and conservation areas, such as lakefronts (Georgian Bay, Lake Simcoe, Kawartha Lakes, part of Lake Ontario) and the Niagara Escarpment.

Other land use policies

As we have seen in earlier chapters, the province assumed responsibility for an increasing number of areas of land use control and planning before and after announcement of the TCR plan. As a result, some of the land use objectives incorporated into the TCR planning documents were actually provincial land use policies which have operated more or less independently of the regional planning program. Examples of objectives which antedated the TCR plan were the policy of restraining urban sprawl and limiting the creation of non-farm residential lots in rural areas and also of the policy of insistence on complete lake-based sewer facilities for new urban subdivisions. Similarly, as the province developed its policy of preserving prime agricultural land, this was added as an objective in the TCR documents. For the purpose of this monograph, we shall distinguish between these general provincial land use objectives, on the one hand, and the objectives of provincial regional planning, on the other. In discussing regional planning, we shall be concerned only with the land use objectives listed on p. 132.

IMPLEMENTATION OF THE TCR PLAN: 1970-8

A review of the extent to which the provincial government implemented the TCR plan during the eight years following its announcement suggests that, contrary to the goals of the plan, the distribution of population and employment within the region in 1978 was not very different from what it would have been in the absence of the provincial regional planning and development program. More important, little had been done by 1978 which would suggest that these programs would have a significant effect on the distribution of economic activity in the year 2001. Few of the actions which would have been required to achieve the TCR goals were taken, and a number of actions which were taken conflicted with the goals.

Having reached this conclusion, however, it is important to add two points. First, the fact that the goals were not pursued relentlessly is not necessarily to be regretted, as two critical reviews of the TCR program have taken for granted.[9] The case for the TCR program on economic efficiency grounds is open to question. Although many of the objectives were perhaps non-economic, the economic costs of the program were not seriously considered by either the politicians or the planners when the program was being adopted.[10] Second, the fact that implementation of the TCR plan has been limited does not imply that the

9 Ontario, Central Ontario Lakeshore Urban Complex Task Force (1974) and Bens, Golden, and Bryant (1977).

10 See Chapter 2.

program has had no effect whatsoever. A number of measures which were influenced by TCR plan considerations are documented below.

Further planning, policy statements, and legislation

Announcement of the 1970 TCR plan was followed by a number of measures which were designed to increase the powers of the province in planning and to translate the general TCR concepts into more detailed land use and development plans for the region and various subregions.

(a) TCR population allocations

In 1971 *A Status Report on the TCR* provided some additional details on population allocations for areas to be serviced by the South Peel and Central York-Durham servicing schemes.[11] In 1972 the TCR Combined Task Forces recommended detailed population allocations for Zone 1 and the Yonge Street corridor,[12] and in 1974 virtually identical population allocations were recommended by the Central Ontario Lakeshore Urban Complex (COLUC) Task Force report.[13] However, neither of these task force reports was adopted as government policy. (See Table 7.1 for details of the population allocations.)

(b) Simcoe-Georgian and Northumberland task forces

Task forces were set up and made recommendations in 1975-6 for plans for the Simcoe-Georgian and Northumberland areas in Zone 3.[14]

(c) Niagara Escarpment

A number of actions were taken to facilitate land use planning and control along the Niagara Escarpment, which passes through the TCR. The Niagara Escarpment Protection Act (1970) and the Pits and Quarries Control Act (1971) strengthened regulations covering pits and quarries. In 1972 the Niagara Escarpment Task Force was established and made recommendations on land use along the escarpment.[15] In a 1973 policy statement on the escarpment the province announced its intention 'to create a planning system featuring strong, provincially-directed land-use regulation plus public ownership where necessary.[16] At the

11 Ontario, Government of Ontario (1971b).

12 Ontario, TCR Combined Task Forces and Regional Development Branch, Ministry of Treasury, Economics and Intergovernmental Affairs (1972).

13 Ontario, Central Ontario Lakeshore Urban Complex Task Force (1974).

14 Ontario, Simcoe-Georgian Area Task Force (1976), and Ontario, Northumberland Area Task Force (1975).

15 Ontario, Niagara Escarpment Task Force (1972).

16 Ontario, Ministry of Treasury, Economics and Intergovernmental Affairs (1973a), p. 9.

same time the province passed the Niagara Escarpment Planning and Development Act, which established a provincially appointed Niagara Escarpment Commission (NEC) to prepare a plan for the escarpment planning area. Once approved by the province, the Niagara Escarpment plan would take precedence over local plans, which would have to be modified to conform with it. The province assumed the power to require any municipality in the area to adopt an official plan. The province specified that the provincial plan would be implemented by development control, that is, that every proposed development would be studied on its merits, rather than by zoning by-laws, and interim development controls were imposed. The interim development controls suspend local zoning by-laws in parts of the escarpment planning area, and in these parts development requires a permit from the NEC. As of 1978, the NEC was still preparing a plan, but the province had acquired 45,135 acres of land in the escarpment area.[17]

At the same time, its critics were saying it had failed.[18] According to Estrin et al., 'as of mid-1977, development control has proved to be a dismal failure. Many environmentally damaging projects have been permitted to proceed.' In 1976 Horton reported:

An extremely high percentage of [escarpment area development] applications have been approved and, more often than not, development permits have been granted in sensitive sections of the control area. The Commission's planning staff has generally recommended approvals and denials in a ratio of 25/75. The Commission members have, in turn, consistently reversed that ratio, increasing the approvals from 25 per cent to 75 per cent. Appeals to Minister of Housing have accounted for approximately 15 per cent additional approvals. Thus, in excess of 85 per cent of all applications for development in Escarpment lands in the area designated for special restrictive controls has been granted. This almost unbelievable situation seems to be explained by two factors, (1) the continued parochialism and limited vision of local officials who see continued development as continued 'progress', and are supported in this view by their appointed representatives on the Commission, and (2) the bias of the Ministry of Housing toward increasing residential development wherever possible in the province – including the environmentally sensitive development control area of the Niagara Escarpment ... The performance of the Commission in its administration of

17 Ontario, Ministry of Natural Resources (1978), p. 18. Horton (1977), pp. 155, 169, reports that between 1968 and 1976 the province acquired approximately 72,000 acres of land in the escarpment area.

18 Bureau of Municipal Research (1977).

TABLE 7.1

Population in 1971 and population allocations for 2001 for TCR Zone 1 and adjacent areas ('000)

Area	1971 population[a]	Population allocations for 2001				
		1970 TCR policy[b]	1971 status report policy[c]	1972 TCR Task Force[d]	1974 COLUC Task Force[e]	1976 TCR[f]
TCR	4,074*	8,000		7,500		
Zone 1	3,083*	5,700	5,700	5,475		
COLUC	3,325				5,647	
Metro and north (Zone 1)	2,165*	3,100		3,285[g]		
Metro and north fringe	2,104			2,815	2,805	
Metro	2,100*			2,750		
North fringe	18*		95[h]	65		
Richmond Hill	25		67[h]	60	59	
Markham-Unionville	11		21[h]	21	21	
Woodbridge	3		20	12	12	
West (Zone 1)	750*	1,850		1,533[g]		
Hamilton	354			600	598	
Burlington	80			135	134	
N. Burlington	1			–	1	
Milton	7	[i]		11	11	
Oakville	57			90	90	
N. Oakville	1			–	1	
Mississauga	143		440[j]	360	356	
Erin Mills-Meadowvale	9		210[j]	186	183	
Brampton-Bramalea	65		275[j]	150	130	
Malton	18		35[j]	30	30	

East (Zone 1)	168*	750		657[g]		
North Pickering	2		75[h]	120	120	
Pickering	21		115[h]	90	90	
Ajax	17			45	45	
Oshawa-Whitby	115			400	397	
Audrey	3			–	3	
Columbus	1			–	33	
Bowmanville	8			35	35	
Zone 2						
Newmarket-Aurora	30		50[h]	75	75	
Georgetown	17			21	31	
Regional Municipality of York						550
Portion of York serviced by Central York Servicing Scheme			238[h]			450[d]
Servicing schemes						
South Peel			950[j]			
Central York Durham			400[h]			

a Figures without asterisks are from Ontario, Central Ontario Lakeshore Urban Complex Task Force (1974). Figures with asterisks are from Ontario, TCR Combined Task Forces and Regional Development Branch, Ministry of Treasury, Economics and Intergovernmental Affairs (1972)
b Ontario, Government of Ontario (1970b)
c Ontario, Government of Ontario (1971a)
d Ontario, TCR Combined Task Forces and Regional Development Branch, Ministry of Treasury, Economics and Intergovernmental Affairs (1972)
e Ontario, Central Ontario Lakeshore Urban Complex Task Force (1974). Figures are for 'Scenario A.' The COLUC report also gives a population allocation for the 'Mature State,' to be achieved between the years 2000 and 2050.
f Ontario, Ministry of Treasury, Economics and Intergovernmental Affairs, Regional Planning Branch (1976a)
g Figures for subregions of Zone 1 may be greater or less than figures for urban areas within those subregions.
h To be serviced by Central York-Durham servicing scheme
i Milton was included in Zone 2 by Ontario, Government of Ontario (1970b) and shifted to Zone 1 by Ontario, Government of Ontario (1971a)
j To be serviced by South Peel servicing scheme.

development control has been regrettable. A much more rigorous and restrictive application of the Commission's guidelines is greatly needed ... The attitudes and actions of the Ministry of Housing have served to establish a climate within which residential development is encouraged. The Ministry has approved new subdivision proposals and continued urban expansion in the planning area – against the strong opposition of the Commission's planning staff, some Commission members, conservation authorities and other interested parties. This has done little to justify or support the government's stated commitment to preserving the Escarpment.[19]

In 1978 the province announced that development control over about 60 per cent of the two million acres then under the NEC would be given back to the area municipalities. Also in 1978, after it turned down a proposal for a major office development in Niagara-on-the-Lake, the NEC reversed its decision under pressure from the municipal and regional governments.[20] In the same year, when the NEC would not allow a hotel and conference complex on escarpment land, its ruling was overturned by the province on the ground that the project would provide jobs.[21]

(d) Parkway belt

In 1973, the Parkway Belt Planning and Development Act provided for the establishment of a Parkway Belt Planning Area west and north of Metro Toronto and applied the provincial planning provisions of the Ontario Planning and Development Act to the area. As a holding device, to stabilize land use within the area until a provincial plan could be prepared by TEIGA and approved by the provincial government and municipal plans and by-laws could be amended, the province temporarily restricted the entire area to agricultural use. A plan was adopted in 1978.

Actions which promoted or contradicted TCR plan objectives

The provincial government has taken actions which appear either to have promoted essential features of the TCR plan, on the one hand, or to be inconsistent with the plan, on the other. We shall now identify these. (We do not, however, list cases of inaction, such as that of the province in not doing anything to extend the parkway belt east of Metro.[22]) The following objectives of the TCR

19 Estrin et al. (1978), p. 338, and Horton (1977), pp. 166–7.

20 *Globe and Mail*, 12 October 1978, p. 7.

21 *Globe and Mail*, 1 November 1978, p. 5.

22 However, the Regional Municipality of Durham has made provision for a parkway belt or something like it in its official plan.

plan are considered: (a) restriction of urban development in the commutershed, (b) the 'go-east' policy, and (c) the parkway belt.

(a) Restriction of urban development in the commutershed

A number of activities appear to have promoted this TCR concept during the 1970s. Even before announcament of the TCR plan, the provincial government was deliberately limiting residential development in rural areas. Announcement of the plan in May 1970 led to further restraint in the short run, in part because uncertainty about the details of the plan (which was drawn up by the former Regional Development Branch of the Department of Treasury and Economics) led to delays in the approval of subdivision plans by the former Community Planning Branch of the Department of Municipal Affairs. According to the Department of Municipal Affairs itself, during 1970 and 1971:

Particular attention was paid to the relationship of proposed plans of subdivision to the Toronto Centred Region Concept ... A number of plans [of subdivision] were not recommended for approval pending the completion of work currently under way on the refinement of the Concept ...

The [Community Planning] branch has committed itself to assist the Department of Treasury and Economics to refining the concept for the Toronto-centred region. This has involved the scrutiny of official plans and plans of subdivision affected by the regional plan.[23]

According to the 1973 report and background papers of the Advisory Task Force on Housing Policy (Comay Report):

In the Central Ontario Region (Toronto-Centred Region), where the Provincial regional planning process has reached the stage of formulating development proposals, the result has been, for all practical purposes, to freeze housing development in critical areas, most notably in the Metropolitan Toronto housing market area. The process has imposed very extended delays on the approval of both municipal and private development plans and discouraged municipalities and developers from proceeding with development plans while regional planning questions remain unsettled. The plan's proposals and guidelines have operated to stall housing development in many serviced residential areas, while failing to provide programs for implementing housing development in other areas.[24]

23 Ontario, Department of Municipal Affairs, *1971 Annual Report*, p. 31.
24 Ontario, Advisory Task Force on Housing Policy (1973c), p. 39.

The Province's regional planning process, particularly in the Toronto-Centred Region, has served to delay both the servicing of residential land and the processing of development plans ... The government's inability to resolve population allocations in the area north of Metropolitan Toronto has forestalled the servicing of major land areas required for the housing needs of the metropolitan area ... The Parkway Belt concept was first announced in May, 1970, and the proposed boundaries for its western section were finally incorporated in legislation in June, 1973. Even now (June 1973), it is still necessary to undergo a process of statutory hearings before the boundaries are finally fixed. It is not unlikely that some proposals will have effectively been held up for a period of at least four years before their owners are finally in a position to proceed with housing development.[25]

Similar views were expressed in 1972 Urban Development Institute studies by Derkowski and FENCO and in a 1973 Ontario Economic Council study.[26] In contrast, some observers have expressed misgivings about how important provincial constraints on development really were.

To cite a specific example of the delays caused by the TCR plan, in April 1970, the month before the provincial government unveiled the plan, the OMB approved an official plan for the Township of Caledon in the Peel area of Zone 2. Since the official plan conflicted with the TCR plan regarding the extent of population growth to be permitted in the commutershed, in September 1970 the OMB agreed to review its approval of the official plan.[27] Punter reports: 'In Caledon the immediate results [of the TCR plan] was the delay of all subdivision approvals of the Community Planning Branch until the [Caledon] Official Plan was amended to conform to the Toronto-Centred Region Plan. This amendment was approved by the OMB on April 30, 1971 ... The confusion created by Official Plan, TCR and Niagara Escarpment Proposals prevented the approval of any subdivisions in the [southern Caledon Township] area until early 1973.'[28]

TCR considerations have also been cited as having played a role in the delay in, modification of, or rejection of several other subdivision and official plan amendment applications in what is now Peel by the Community Planning Branch in the early 1970s: Meadowvale North, North Dixie, amendments 30 to 32 of the official plan for Chinguacousy Township, and amendment 3 of the Toronto

25 Ontario, Advisory Task Force on Housing Policy (1973b), pp. 15–16.
26 Derkowski (1972); Urban Development Institute, Region of York Committee (1972); and Ontario, Ontario Economic Council (1973), p. 88.
27 *Ontario Municipal Board Reports* 2, pp. 1–5.
28 Punter (1974), p. 198.

Gore official plan. However, there is not a consensus on the role the TCR plan played in these matters.

The provincial government stated in 1971 that three privately sponsored urban development projects on the new town scale in Zone 2 collapsed or were not approved because of conflict with the TCR plan: Century City, a Revenue Properties project for 50,000 people in Uxbridge Township near Stouffville; Centennial City, a Milani and Milani Holdings project for 50,000 people north of Woodbridge; and Castlemore, a project for 25,000 people in Toronto Gore Township. Assembly of 6,000 acres of land for Century City had begun in 1968 and a draft plan of subdivision had been submitted to the provincial government in 1970. Castlemore did not reach the formal application stage.[29]

The activities did not end in the early 1970s, for as late as 1977, in refusing to approve an amendment to the Town of Vaughan official plan which would have permitted a large residential development in the commutershed near Kleinburg north of Toronto, the OMB stated:

In examining the need for housing in the area of Kleinburg, the Board must have regard for the policies and intentions expressed by the region, the provincial Government and other authorities as well as specific requirements for this area ... The Toronto-Centred Region Plan ... provides for a Greenbelt across the top of Metro Toronto, referred to as the commutershed, Zone 2. The policy for this Zone 2, is to retain it to the maximum degree in recreational, agricultural and other open space uses ... The growth which is expected to take place in Zone 2, is to be encouraged into the vicinity of the urban areas along the Yonge Street corridor and in specific communities, not including Kleinburg.[30]

However, there are three ways in which the province retreated from the initial TCR plan. First, in 1971 the province announced that Zone 1 would be expanded at the expense of Zone 2, thus permitting urban development in three areas where such development was initially to have been restricted: in the Hamilton area, in and near Milton, and adjacent to Brampton.[31]

29 Ontario, Government of Ontario (1971a), p. 14; Martin and Matthews (1977); Ontario, TCR Combined Task Forces and Regional Development Branch, Ministry of Treasury, Economics and Intergovernmental Affairs (1972), p. A8; and Martin (1975), p. 24. It should be noted that these three proposed developments violated provincial land use policies which antedated the TCR plan, e.g., servicing requirements. However, evidently the TCR plan clarified these policies, since it is doubtful that the developers would have proceeded with the land assemblies if the policies had been clear.

30 *Re Vaughan Planning Area Official Plan Amendment 19*, in *Ontario Municipal Board Reports*, June 1977, pp. 327–35.

31 Ontario, Government of Ontario (1971a), pp. 8–9.

Second, in 1971 the province announced that the population of the portion of the Regional Municipality of York to be served by the Central York-Durham servicing scheme would be restricted to about 238,000 in the year 2000.[32] The Regional Municipality of York argued for a larger population allocation, and a figure of 416,000 was accepted by the province in 1973 as a basis for the design of the servicing scheme.[33] By 1976 this figure had been revised upward to 450,000.[34]

Third, the province introduced a third GO transit commuter rail line to Georgetown in Zone 2 northwest of Metro Toronto. To the extent that this would stimulate development along a northwest corridor, it would make it more difficult to achieve the Zone 2, linear urban system, and go-east objectives of the TCR plan.

(b) Go-east

Three decisions by the province support the 'go-east' objectives of the TCR plan. First, the province favoured the Pickering site northeast of Metro Toronto for the federal government's proposed second Toronto international airport, which was announced in 1972 (and which has since been shelved), and stated that, in addition to other considerations, 'the location of the airport east of Toronto is the result of joint Federal-Provincial effort to provide a major stimulus to development east of Metropolitan Toronto, as called for in the Toronto-Centred Region Plan.'[35] Simultaneously the province announced plans to develop a provincially sponsored new town at North Pickering, south of the proposed airport and approximately 18 miles northeast of central Toronto, and the province proceeded with land acquisition.[36] In 1975 the province established the North Pickering Development Corporation to plan and develop a new community over a 15 to 20 year period as 'part of provincial policy to stimulate economic growth in the area east of Metro Toronto, and to relieve growth pressures in the areas immediately north and west of Metro.'[37] As of 1978 it was not clear how much development would actually take place in North Pickering. It should be noted, however, that some of those involved in planning for the TCR felt that

32 Ibid.
33 Ontario, Central Ontario Lakeshore Urban Complex Task Force (1974), p. 16.
34 Ontario, Ministry of Treasury, Economics and Intergovernmental Affairs, Regional Planning Branch (1976c).
35 Ontario, Commission of Inquiry into the Acquisition by the Ministry of Housing of Certain Lands in the Community of North Pickering (1978), p. 3.
36 Ontario, Ministry of Housing (1975) and Ontario, Ontario Land Corporation (1977/78).
37 Ontario, Ministry of Housing, *Annual Report 1976/77*, p. 14.

development of North Pickering would detract from the development of Ajax and Oshawa, and hence did not support the basic go-east policy.[38]

Second, the province is constructing and heavily subsidizing the Central York-Durham servicing scheme, which will service future development east of Metro Toronto in Durham. However, it will also support development north of Metro, and to a larger extent than to the east.

Third, in 1977 the province decided to relocate the Ministry of Revenue to Oshawa and the Liquor Control Board of Ontario warehousing operation to Whitby.

However, the government did not restrain or refrain from encouraging development west or north of Metro Toronto or in Metro itself in order to push development east, as some of the planners recommended.[39] Apart from the large revised population allocation for the Regional Municipality of York and the GO train to Georgetown, the Ontario Housing Action Program (1973-8) was designed primarily to accelerate residential development in a number of areas near Metro Toronto, including the regional municipalities of Peel, York, and Durham. Because the major servicing schemes were installed in Peel prior to OHAP whereas those in Durham were being installed while OHAP was in effect, OHAP operated rather differently in Peel and Durham. In Peel, since services were already available, OHAP emphasized approval of land for development and capital housing incentives grants. In Durham, OHAP emphasized provision of interest-free loans for water and sewer systems. It is evident, however, that the province pursued its housing objectives west of Toronto in Peel without being constrained by the TCR plan.

Subject to Approval pointed out in 1973 that 'the popular belief that the TCR plan is aimed at the decentralization of economic activity or any other effective limitation of metropolitan growth is placed in serious question by the government's explicit support of the Metro Centre proposals, not to mention its own rather massive building program in central Toronto.'[40]

(c) Parkway belt

The province has moved ahead to create a parkway belt west and north of Metro Toronto. However, the parkway belt evolved into something less than was

38 Ontario, TCR Combined Task Forces and Regional Development Branch, Ministry of Treasury, Economics and Intergovernmental Affairs (1972), p. 123, and Ontario, Central Ontario Lakeshore Urban Complex Task Force (1974), p. 48. See also Bossons (1978), p. 148n.

39 Ontario, TCR Combined Task Forces and Regional Development Branch, Ministry of Treasury, Economics and Intergovernmental Affairs (1972).

40 Ontario, Ontario Economic Council (1973), p. 22.

originally suggested. The map accompanying the 1970 TCR plan suggested that the east-west portion of the parkway belt would be about one mile wide (see Figure 7.2), and in 1971 the province stated that the parkway belt would have four functions: utility corridor, urban separation to define identifiable communities and prevent continuous sprawl, environmental buffer for the transportation system, and land reserve for open space, recreation, and low-intensity uses.[41] When the Parkway Belt West Planning Area was created in 1973, the east-west portion of the planning area was generally one-half mile wide or less; the Mississauga portions, which accounted for about one-third of the length of this belt, were only one-tenth to one-fifth mile wide.[42] The 'Hole in the Donut' area west of the Malton airport was not included in the parkway belt. Thus, over much of its width the parkway belt would inevitably be simply a utility corridor. Moreover, after 1973 the belt continued to shrink; for example, the 1977 report of the Hearing Officers on the 1976 Draft Plan recommended that a large number of areas be deleted from the belt and that more intensive land uses including urban uses be permitted on substantial areas retained in the belt.[43] Thus, although the parkway belt may be significant as a measure to plan for utility corridors in advance of demand, there is some doubt whether the belt will have an important effect on the structure of the Toronto-centred urban system or the regional distribution of economic activity. This statement conflicts with the government's own assessment in 1976, when it was stated that 'The Parkway Belt System ... is intended to perform a number of functions. Primarily, the system places open space between built-up areas. In addition, in some places it will serve as a corridor for transportation and communication facilities.'[44]

In the Parkway Belt West Plan, which was given final approval in 1978, the parkway belt west has an area of 52,000 acres. This includes 8,000 acres of public parkland, 22,000 acres of publicly owned utility corridors, and 22,000 acres of privately owned land subject to various development restrictions. As of 1978, the government had acquired about 21,000 acres of land at a cost of $192 million, and it was estimated that the remaining 9,000 acres designated for public acquisition would cost an additional $175 million in 1978 dollars.[45]

41 Ontario, Government of Ontario (1971a), p. 7.
42 Vito (1973).
43 Ontario, Hearing Officers – Parkway Belt West (1977).
44 Ontario, Ministry of Treasury, Economics and Intergovernmental Affairs, Regional Planning Branch (1976c), p. 9.
45 *Globe and Mail*, 27 July 1978, p. 3.

CURRENT PROVINCIAL REGIONAL POLICY

If one reads through provincial policy statements and planning documents issued during the period since 1970, one finds that lip service has been paid to the TCR plan throughout the period. As recently as 1976 the government stated: 'The key objectives of the TCR plan remain valid today, and major steps have been taken during the last five years to put it into effect. The policy guides the municipalities in the region in preparing their official plans and the government in coordinating and directing the activities of its ministries and agencies in support of the development concept. As a result, development in the Toronto-Centred Region since 1971 has been generally consistent with the policy.'[46]

Nevertheless, the attitude of the government toward the plan and toward provincial regional planning has changed considerably over this period. Indeed, the province has backed away from comprehensive regional and province-wide planning of the type suggested by the initial Design for Development documents. This conclusion is supported by the discussion above on the implementation of the TCR plan, which indicates that government intervention on behalf of the plan was, in fact, limited. This conclusion is further supported by several other observations. First, although the province adopted the TCR plan and Northwestern Ontario strategy as government policy during 1971,[47] it did not adopt plans for the other six regions of Ontario or for the province as a whole.

Second, after adopting the TCR plan and providing some sketchy elaborations during 1971, the province did not proceed to fill in details of the plan on matters such as population allocations, phasing, and techniques of implementation. Although task forces were set up for this purpose, the province did not adopt the recommendations of either the TCR Combined Task Forces in 1972 or the COLUC Task Force in 1974.

Third, after changing the title of the Regional Development Branch of TEIGA to the Regional Planning Branch in 1973, in 1976 the province moved away from both the 'regional' and 'planning' emphasis by replacing the Regional Planning Branch by the Economic Development Branch. One implication was that the province wished to decentralize responsibilities among the provincial ministries rather than concentrate them in TEIGA. Another implication was that the province wished to downplay comprehensive regional land use planning and

46 Ontario, Ministry of Treasury, Economics and Intergovernmental Affairs, Regional Development Branch (1976c), p. 3.
47 Ontario, Government of Ontario (1971a, b).

control and focus instead on the problem of disparities in the level of economic development across the province. The emphasis is on promotion of development in areas of slow growth rather than on restraint of development in areas of rapid growth. This is consistent with the Ontario Regional Priority Budget, which has been used since 1973 to fund socioeconomic projects in the north, with the establishment of the Ministry of Northern Affairs in 1977, and with the publication in 1977 and 1978 of two reports which state that the province remains committed to promoting the development of northwestern Ontario.[48] Within the Toronto-centred region, attention has recently been focused primarily on areas where the province wishes to promote economic development, particularly the Simcoe-Georgian, Northumberland, and Durham areas, e.g., support for the Barrie annexation scheme, the decision to relocate the Ministry of Revenue and the LCBO warehousing operation east of Metro Toronto, and subsidies for servicing and an industrial park in Durham.

In short, although the province's regional planning program has been characterized by elaborate administrative reorganizations,[49] there has been little in the way of concrete regional policy, particularly outside the TCR.

The explanation for the limited scale on which the TCR plan has been implemented and for the retreat of the province from the comprehensive regional planning program is not difficult to understand. First, as we have indicated, one suspects that the government was never entirely serious about its intention to carry out what it announced to be 'policy' for the TCR. Second, it appears clear that before the plan was announced the planners and politicians had failed to consider the scale of government intervention and the substantial sacrifice of non-TCR plan objectives that would be required to achieve the TCR plan objectives. Any attempt to implement the plan ran up against high costs from a number of points of view: the aggregate opportunity costs in terms of such things as costs of production of goods and services and the level of consumption of housing services; the cost to the taxpayer of land acquisition, subsidies, and provision of infrastructure ahead of demand; and the political costs of dealing with would-be homeowners who could not afford a house, property owners whose land would be reduced in value or expropriated, and local and regional municipalities whose plans for growth conflicted with provincial policy.

For example, the political problems that arose in connection with the land acquisition programs for the Niagara Escarpment (Forks of the Credit) and the

48 Ontario, Ministry of Northern Affairs (1977), and Ontario, Ministry of Treasury, Economics and Intergovernmental Affairs (1978).

49 Tindal (1973).

provincial new towns, land use controls in the Niagara Escarpment planning area and the Parkway Belt West, the regional government program, and the proposal for restraint of development in the Regional Municipality of York reduced the interest of the government in regional land use planning. Evidently the government concluded that its poor showing in the 1975 election was due in part to its land use and regional policies, and this conclusion encouraged the retreat from regional planning.

The 1972 and 1974 task force reports, which represented the views of planners, could not be adopted because they called upon the government to accept these costs and commit itself to implement the TCR plan in spite of them. For example, in 1972 the planners asked the government to accept the need for provincial restraints on development in Zone 1 west and particularly northwest of Metro Toronto in what is now the Regional Municipality of Peel. The 1972 report stated that with the exception of Erin Mills, in Peel 'growth should be controlled' and 'further growth in this area [Brampton-Bramalea-Malton-Woodbridge-Markham-Unionville], more so than anywhere else, could lead to a complete breakdown of the TCR concept ... The growth of these urban places must be strictly controlled.'[50] The province was unwilling to exercise such control over development in the west.

In 1973 the province indicated that land acquisition programs along the escarpment and Parkway Belt West would cost the taxpayer \$250-\$500 million and \$150-\$200 million respectively,[51] and by 1975 the figure for the Parkway Belt West had risen to over \$500 million.[52]

A third explanation for the shift in provincial regional policy was the economic climate in the mid-1970s. Of particular importance was the shortage of serviced residential land approved for development and the escalation of housing prices during the early 1970s. The TCR plan completely ignored the provision of housing among its objectives, but by 1973-4 and the Comay Report, the new Ministry of Housing, and the Ontario Housing Action Program, the province was giving creation of building lots priority over implementation of the plan. Since the Peel sewer system serving the area west of Toronto was in place whereas the Durham system serving the east was not yet ready, accelerated housing development was inevitably concentrated in the Bramalea and Erin Mills areas of Peel.

50 Ontario, TCR Combined Task Forces and Regional Development Branch, Ministry of Treasury, Economics and Intergovernmental Affairs (1972), pp. 32 and 35–6.

51 Ontario, Ministry of Treasury, Economics and Intergovernmental Affairs (1973a), p. 31, (1973b), p. 33.

52 Ontario, Ministry of Treasury, Economics and Intergovernmental Affairs, *1975 Budget*, p. 16.

Also, the high unemployment rate in the mid-1970s made the government unwilling to control the location of employment. In short, development of housing and jobs took priority over the TCR plan.

ECONOMIC ANALYSIS OF PROVINCIAL REGIONAL PLANNING

Economic rationale for the Design for Development program

We have already discussed at some length in Chapter 2 the lack of careful economic analysis, and particularly the neglect of opportunity costs, in the province's regional planning program and TCR plan.

Any economic analysis of proposed provincial intervention in land markets of the type contemplated by the Design for Development program should begin with identification of potential market failures which would provide a possible justification for government activities. At the outset it must be recognized that many forms of government intervention in land markets have been associated with the Design for Development program. The core of the program was the objective of bringing about a size and spatial distribution of urban areas different from what the market would produce. However, other aspects of the program involved preservation of land for open space, recreation, and future interurban transportation and utility services.

The case for government action to set aside land for future interurban transportation and utility corridors is completely straightforward. Since the government provides interurban transportation services, etc., obviously demands for land for future services will not be represented in the private market, and the government must act to designate and acquire land in advance for such projects. Moreover, since the future land use patterns and transportation and servicing schemes are interdependent, obviously there will be significant gains in terms of reduced uncertainty if the government can plan for such schemes in advance and make its plans known to the public. Thus, to a large extent activities like the provincial planning and land use controls in the Parkway Belt West, which is largely a transportation-utility corridor, may be justified by these considerations. However, although this explanation provides a possible justification for such activities, it should be emphasized that benefit-cost analysis should be used explicitly in such projects. To date the province has failed to carry out and make public benefit-cost analyses for its interurban transportation and servicing schemes.

The case for government action to set aside land for future public recreational use and public open space based on public goods considerations is also straightforward. This would provide a potential justification for activities such as development restrictions on lakeshores or on the Niagara Escarpment, but obviously in the absence of cost-benefit analyses one has no idea whether the actions the province has actually taken in these areas have increased the efficiency of resource allocation. Much depends also on whether the government not only restricts private development but takes steps necessary to secure public access to the areas in question or to develop them for public use.

Finally, the case for government action to change the size and spatial distribution of urban areas rests primarily on the macro-urban externalities discussed in Chapter 2 (pp. 22-3). As we indicated there, the empirical analysis of macro-urban externalities is so limited that at present the case for provincial intervention on these grounds is very uncertain. What is clear is that the rationale provided in Design for Development documents for provincial intervention in these matters does not give the slightest evidence that such intervention would increase the efficiency of resource allocation. The idea that the size and spatial distribution of urban areas should be determined simply to minimize the cost of certain public services such as waste disposal, to minimize congestion, or to create 'identifiable communities,' etc., is absurd, since this completely neglects opportunity costs. It should also be noted that policies to decentralize development would generally lead to an increase in rate of conversion of land from agricultural to urban use, since larger metropolitan areas use less land per capita than smaller urban areas, and hence would conflict with other provincial land use objectives.

Effect of the TCR Plan on land prices

Whatever the economic rationale for and merits of the Design for Development program and TCR plan, it is of some interest to consider the effect of these provincial activities on the supply and price of residential lots and houses, both during the land and housing boom in the early 1970s and in the longer term.

It appears that provincial planning in the TCR (or, more accurately, the temporary freeze it caused in the development approval process) made a relatively minor short-term contribution to the escalation of serviced residential land and housing prices during the first half of the 1970s. However, by the second half of the 1970s the TCR planning effort had more or less disappeared (apart from some residual effects on local official plans and, of course, the Parkway Belt West and Niagara Escarpment planning programs), and hence it no longer had, and in the foreseeable future would not have, any significant effect on the supply or price of serviced residential land.

The conclusion that the effect of provincial regional planning on serviced land and housing prices was minor is supported by two principal findings. First, the major factors responsible for the rise in prices during the boom in the first half of the 1970s were on the demand side of the market.[53] Second, the regional planning program and TCR plan were implemented to only a limited extent. Measures undertaken by the province to support the TCR objectives were less important than a number of other types of government activity, such as the normal subdivision approval process (which would have been complex and involved many intentional and unintentional restrictions even without provincial regional planning), municipal resistance to development, and the limited provision of central lake-based sewage capacity combined with restrictions on other forms of sewage disposal.

Nevertheless, it does appear that announcement of the TCR plan in 1970 reduced the rate of subdivision and consent approval by the province during 1970-3. First, a few large urban development projects north of Metro Toronto evidently collapsed or were rejected in part because of conflicts with plan objectives. Second, for about three years after the announcement of the plan, approvals for subdivisions in various areas were held up because of uncertainty about details of the plan, such as the location of the parkway belt and the York Region population allocation, and the need to revise official plans. Furthermore, provincial planning for the TCR further increased the complexity of the subdivision approval process. It thus appears that, in the absence of provincial TCR planning, the rate of subdivision and consent approval in the early 1970s and the rate of housing construction in the first half of the 1970s would have been marginally higher than they were.

It should be added, however, that if the TCR plan had been vigorously implemented, it would have had a substantial and long-term effect on the price of serviced residential land and housing in the Toronto metropolitan area. Some of the planners involved in the TCR program evidently believed that, by diverting employment and population away from the area, the plan would lead to a reduction in housing prices in Toronto as well as lower housing prices for the people who were diverted from Toronto to smaller centres.[54] In so far as the decentralization objective is concerned, this result would be plausible if the only methods used to implement the TCR were controls on employment in the Toronto metropolitan area and subsidies for employment elsewhere, but it would not be plausible if restrictions on residential development within commuting distance of Toronto were used, as the planners evidently intended.

53 Scheffman (1977) and Canada, Federal/Provincial Task Force on the Supply and Price of Serviced Residential Land (1978).

54 Ontario, TCR Combined Task Forces and Regional Development Branch, Ministry of Treasury, Economics and Intergovernmental Affairs (1972), p. 29.

The fact that such restrictions were considered indispensable to achievement of the TCR objectives is indicated by the 1974 COLUC Task Force Report. According to the report, in order to encourage development to the east of Metro Toronto:

> measures to control the pace of development to the north and west will be essential. The need for firm steps to maintain a ceiling on the Yonge Street corridor has already been discussed. But unrestrained growth to the maturity level between Toronto and Hamilton would also frustrate the eastern corridor policy. Rationing of growth to maintain a tight land market for both industrial and residential development in the west will help to deflect such development to the east. Yet such restraints appear to be directly opposed to current government policy, and specifically the Housing Action Program ... Stringent rationing of development land would inevitably tend to raise residential land prices in the west ... In view of the undoubted magnitude of the task involved in implementing the 'go-east' policy, and the apparent conflict with current policies relating to housing, some members of the task force question the feasibility of the policy.[55]

Moreover, for any given population level in the Toronto metropolitan area, achievement of the other objectives – creation of a parkway belt with substantial open space, restriction of development in Zone 2 and confinement of development to a corridor along the lake, and preservation of prime agricultural land – would obviously lead to higher land and housing prices at any given distance from the centre of Metro Toronto. Withdrawal of land from residential development would lead to an expansion of the urban boundary where development was permitted and a higher location premium for land at any given distance from the centre.

In fact, the adverse effect which vigorous implementation of the TCR plan would have had on housing prices was one of the considerations which led the provincial government not to implement the plan more than it did. In 1973 the government initiated the Ontario Housing Action Program to increase the supply of new dwellings. OHAP offered incentives to municipalities to speed up approval of development applications and open serviced land to early development in a number of 'housing action areas' (HAAs), even though this conflicted with TCR plan objectives. According to the 1974 COLUC Task Force report: 'Rapid development in certain HAA's might well lead to population levels substantially

55 Ontario, Central Ontario Lakeshore Urban Complex Task Force (1974), pp. 17–18.

exceeding COLUC phasing preferences. Since Cabinet has directed that OHAP is to be regarded as a priority program and is not to be inhibited by COLUC preferences, the possibility exists that a serious distortion of the distribution of population envisaged in this report could result.'[56]

56 Ibid., p. 18.

8

Implications for public policy

We shall now bring together this study's principal conclusions and suggestions for provincial land use policy, without providing a summary of the entire study. We hope that people who choose to begin by reading the present chapter will be persuaded to delve into the preceding chapters for more complete discussions of the issues considered here.

THE NEED FOR ECONOMIC ANALYSIS

Any economist who reads much of the literature relating to provincial land use policy in Ontario will be impressed by an obvious paradox. On the one hand, the essential issues at stake in land use policy are traditional economic ones, involving determination of the forms of government intervention in the economy which are required in order to bring about a socially optimal allocation of scarce resources. The methodology of economics has been developed specifically to deal with this problem. Put simply, this is what economics is all about. Yet, there is an almost complete absence of economic analysis in the literature on provincial land use policy. This is true of provincial policy statements, OMB decisions, government-sponsored reports prepared by both government departments and private consulting firms, and independent studies carried out by organizations such as the Bureau of Municipal Research and by academic geographers and planners. In the literature on provincial land use policy, economic analysis has been displaced almost entirely by the conventional wisdom of planners.

The essential weakness of the conventional wisdom of the planning profession is that it advocates the pursuit of a particular set of goals without concern for the opportunity costs. For example, a typical planning document will recommend that all land with a high capability for agriculture should be preserved for agricultural use. Planners seem to find the validity of such statements to be self-evident, although common sense and the methodology of economics suggest that whether a particular parcel of land should be preserved for agriculture depends on the social value of its use in agriculture compared to the social value of its use for other purposes, that is, the social benefits and opportunity costs of preserving it for agriculture. The methodology of economics does *not* insist that the social values of land in alternative uses must be based on market prices alone, but it does insist that the social values in different uses must be compared. Failure to do so will lead to policy decisions which result in wasting of scarce resources. Adherence to the ad hoc 'principles of good planning' is therefore irrational and irresponsible as a foundation for provincial land use policy.

In this study we have repeatedly stressed the importance and usefulness of economic analysis, and particularly explicit empirical cost-benefit calculations, in the formulation and evaluation of provincial land use policy. It should be noted that governmental use of cost-benefit analysis is more widespread in both the United Kingdom and the United States than in Ontario. We hope that the recent increased concern on the part of the federal government with the economic impact of policies will lead to a much more significant role for cost-benefit analysis at the provincial and local levels in the formulation and evaluation of land use policy in Ontario.

THE NEED FOR EMPIRICAL RESEARCH

We have explained in Chapter 2 that Adam Smith's Invisible Hand Theorem states that in the absence of market imperfections resources will be allocated efficiently (in the Pareto sense) by private markets without government intervention. To justify government intervention on the grounds that it would lead to a more efficient allocation of resources, one must establish *empirically* that there are market imperfections which lead to significant market failures, and one must assess *empirically* the benefits and costs of the proposed form of government intervention.

One of the serious impediments to evaluating existing and proposed provincial land use policies is that little of the empirical research required for an economic evaluation of these policies exists. For example, the alleged market imperfections which have been suggested by planners and agricultural economists as a justification for provincial restrictions on rural non-farm residential development involve (among other things) various externalities between farm and non-farm land uses. However, there is no systematic or quantitative information on the extent of these alleged externalities, or whether they are ameliorated by private institutional arrangements, or whether they could be dealt with by government policies other than provincial land use controls. Similarly, there are no studies of people's willingness to pay for the public goods characteristics of farmland as open space, and there are no studies which permit one to draw conclusions about whether Toronto will exceed the optimal city size in the absence of government intervention.

Because of this lack, one of our principal conclusions is that there is at present no empirical justification on efficiency grounds for a number of the major provincial land use planning activities, such as restriction of rural non farm residential development, restriction of conversion of agricultural land, or policies to change the size and spatial distribution of urban areas in the Toronto-centred region. This fact generally does not bother planners, many of whom evidently believe that matters of land use are too important to be left to the private market anyway. However, we have argued that the burden of proof should be on the government when it decides to intervene in the allocation of resources.

Of course, we recognize that many people, including conservationists, environmentalists, and many academic geographers and planners, would argue that the provincial government is not in fact sufficiently motivated to intervene in the allocation of land: to preserve wilderness areas, the Niagara escarpment, prime agricultural land and the fruitlands, and so on. In the absence of empirical analysis, one cannot determine the merits of this argument. The possibility is obviously of considerable potential importance, however, and our recommendation that the burden of proof should be on the government when it decides to intervene in the allocation of resources would not have any practical effect in such situations. Consequently, we also recommend that the government regularly undertake or sponsor, on a greater scale than at present, empirical economic evaluations of important policy areas (e.g., policies regarding parklands and conservation areas, cottage areas, agricultural subsidies) to determine, among other things, whether additional intervention would be justified on economic grounds.

ECONOMIC IMPACT STATEMENTS

Based on the considerations just discussed, we recommend that the provincial government commit itself to a formal practice of preparing economic impact statements in the process of evaluating alternative policy options. These statements would provide an assessment of the economic benefits and costs of proposed undertakings and their distributions among relevant population groups, and they would be prepared before the government decided on a course of action. In order to make sure that such a system would in fact be implemented, we suggest that a provincial agency such as the Ministry of Treasury and Economics be given the responsibility of making sure that the other ministries carry out appropriate economic analyses of proposed policies in their respective areas.

In addition, in the course of revising the Planning Act and similar pieces of legislation, the provincial government should include provisions which would give a more explicit role to economic analysis. For example, the government should take the steps necessary to ensure that in the future the Ontario Municipal Board will give explicit consideration to economic benefits and costs in its decisions.

PERSONNEL POLICY

From a practical point of view, it is important to recognize that in order to implement the recommendations made above the provincial government will require more personnel who have been trained in economic analysis. Thus, in hiring people for the civil service or to work as consultants, and in appointing people in bodies such as the Ontario Municipal Board, more weight should be given to competence in economic analysis (and less to alternative backgrounds such as traditional legal and planning educations).

One might hope that one by-product of such a modification in provincial hiring policy would be to encourage schools of planning to modify their curriculums to ensure that people receiving degrees in planning are trained in basic economic analysis.

POLICY STATEMENTS AND CONSISTENCY OF ACTION

One of the most notable characteristics of provincial intervention in land markets is the extent to which the province has failed to provide clear statements of its policies. The lack of clear statements has resulted in considerable uncertainty among developers, municipal planners, and even provincial representatives concerning what provincial policy is, and it has given the OMB an inordinate amount of power to determine land use policy by default. We recommend that the province provide clear policy statements for use by the OMB as well as municipal governments and landowners.

Another thing which we have observed repeatedly in examining provincial land use policies is that there is a significant discrepancy between what the government claims its policy is and what it has actually done to influence the allocation of land in the province. Thus, although provincial restrictions on rural non-farm residential development have limited the extent of such development and raised the prices of rural lots, there continues to be a significant amount of rural non-farm residential development which violates provincial Urban Development in Rural Areas policies. Similarly, there continues to be development along the Niagara Escarpment which conflicts with the province's stated commitment to preserve the escarpment. Upon examination, it is not clear that provincial intervention in the Niagara urban boundaries issue will save a significant amount of tender fruitland during the next two decades. And in the case of the Toronto-centred region plan, the province did little to redirect urban growth into a two-tier linear corridor or to redirect growth east of Metro Toronto.

A certain amount of obfuscation is an inevitable result of the political process, even though it increases the uncertainty and complexity of the development process and hence increases the cost of allocating resources. It also undermines the credibility of the government. We recommend that the government should consistently implement the policies that it has proclaimed. When the government decides that it is no longer desirable to implement a policy, that policy should be explicitly modified or revoked.

PROVINCIAL GUIDELINES FOR MUNICIPAL DEVELOPMENT CONTROLS

Virtually all commentators on government regulation of the development process in Ontario, including the recent Planning Act Review Committee, have advocated a simplification of such regulation and, in particular, an increase in the autonomy of municipalities. The current system obscures both political accountability for development decisions and the criteria by which the appropriateness of local decisions is judged, and it also increases uncertainties and complexities, and therefore the costs of the development process. The increase in development costs is naturally reflected in higher housing prices. We concur with the recommendation that municipalities should be given the authority to implement their development control policies within a set of basic provincial guidelines. We would urge, however, that provincial guidelines be formulated and implemented and appeals be considered within a formal framework that assesses the economic impact of any proposed action.

We differ with the Planning Act Review Committee's recommendations on provincial guidelines on development standards, however. The Planning Act Review Committee has recommended that, to the extent possible, development standards should be uniform across the province and that the power of municipalities to 'sell' development rights to developers and to use fiscal zoning should be severely limited. The motivation behind these proposals is a concern with equity. We conclude that these recommendations may have an adverse effect on the efficiency of resource allocation. On the basis of our analysis of local development controls presented in Chapter 4, we suggest that efficient resource allocation in urban land and housing markets requires fiscal zoning, and efficiency may be facilitated by a municipality's ability to 'sell' development rights. Furthermore, the level of restrictiveness of municipal development controls required for efficient resource allocation will differ among municipalities. Therefore, we would argue that imposition of development standards by the province cannot be justified without a thorough assessment of the costs and benefits. The efficiency losses resulting from uniform provincial development standards and restrictions on the power of municipalities to sell development rights and impose fiscal zoning must be balanced against the (possible) increase in equity presumed to result from such a policy.

RURAL AND AGRICULTURAL LAND USE IN ONTARIO

Much of the public debate over rural land use policies has taken place without the benefit of the available facts concerning the patterns and trends in rural and agricultural land use. We have provided the first comprehensive survey of the data on this subject, which were previously available only in widely scattered studies.

The data on land conversion indicate clearly that in the aggregate the rate of conversion of agricultural land to built-up urban use is low in relation to the rate of productivity increase in agriculture, the stock of agricultural land, and the decrease in the acreage of census farms. During the past decade about one per cent of the good agricultural land in the province was converted to built-up urban use, and rural-urban land conversion amounted to only about ten per cent of the decrease in the area of census farms.

The data also indicate that in the aggregate the rate of conversion of land to active rural non-farm residential use has been even lower than the rate of conversion to built-up urban use. Furthermore, the land converted to non-farm residential use has come in disproportionately large amounts from lower-quality agricultural land rather than from prime agricultural land.

These findings indicate that there is little basis in fact for the cataclysmic rhetoric which has sometimes characterized recent discussions of the agricultural land issue.

RURAL NON-FARM RESIDENTIAL DEVELOPMENT

Our examination of non-farm residential development in rural areas leads us to the conclusion that general provincial policies designed to restrict the amount of such development are inappropriate. If non-farm residential development in rural areas leads to municipal restrictions on efficient farm operations which non-farm residents find annoying, then the appropriate provincial response is to limit the power of municipalities to restrict *efficient* farm operations rather than to restrict the pattern of residential development in the province. Of course, some farm operations which rural non-farm residents object to may in fact be inefficient (in the Pareto sense), and the power of municipalities to restrict such operations should not be limited by the province. Obviously, before the province can act on our recommendation in this matter it will have to devote resources to determining which of the farm operations which bother non-farm residents are efficient and which are not.

Some observers argue that conflicts between agricultural and non-farm land uses are too pervasive to be handled simply by restrictions of the type indicated above and by other measures such as weed control by-laws and the Agricultural Code of Practice. We have seen no evidence to back their argument but the matter cannot be settled without further empirical study. Pending such a study, there is no justification for provincial intervention on the basis of this argument, as some agricultural economists and rural planners have recommended (e.g., James F. MacLaren, Ltd. 1975). However, in the event that the argument is supported by empirical analysis, it should be stressed that it would not provide a justification for restrictions on the total amount of rural non-farm development but only for zoning-type controls to separate farm and non-farm land uses.

PRIME AND UNIQUE AGRICULTURAL LAND

Many people seem to believe that all of the better agricultural land should be preserved for present and future agricultural use and that urban and other non-farm development should be restricted to inferior (e.g., classes 5 to 7) agricultural soils. In fact provincial land use guidelines have adopted this basic approach. There are two serious weaknesses involved in this naïve policy prescription. The first is that any rational policy toward preservation of agricultural land cannot possibly leave the aggregate quantity of land preserved to be determined by supply factors alone without regard to demand conditions; there is no reason to believe that the total supply of classes 1 to 4 agricultural land and unique soils corresponds in any particular way to the socially optimal amount of land which should be preserved for agriculture.

The second weakness of the naïve policy prescription that the better agricultural land should be preserved for agriculture is that it ignores the fact that the opportunity cost (value in non-farm use) per acre of preserving different parcels of land for agriculture varies. There are two principal reasons for the variation.

First, many of the physical characteristics (slope, drainage, water table, etc.) which make certain parcels of land have a low capability for agricultural use, or which raise the capital and operating costs per unit of agricultural output, also raise the costs of using these parcels for urban use. Second, while location per se (at least within southern Ontario) is of no significant concern for most agricultural uses, location is a principal determinant of the value of different parcels of land for non-farm use. For example, land located five miles from the centre of a large urban area sells for several times as much per acre for non-farm use as otherwise identical land located in a distant rural area. The price differential between otherwise identical land in different locations largely reflects the capitalized value of the difference in transportation costs (e.g., commuting costs) for the parcels.

In dealing with the heterogeneity of land, provincial land use guidelines in the *Green Paper on Planning for Agriculture* have indicated that prime agricultural land should be preserved for agriculture and urban development should generally be channelled into areas with lower potential for agriculture. There is, in fact, no prima facie case for such guidelines, because they ignore the fact that the opportunity cost per acre of preserving different parcels of land for agriculture varies. Any decision to preserve a particular area for agriculture should be based on an explicit consideration of the costs as well as the benefits. An integral part of provincial guidelines governing the conversion of prime and unique agricultural land should be an explicit determination of the costs the province feels 'society' is willing to incur to preserve an acre of such land. Within this framework municipalities could determine whether proposed actions are consistent with provincial policy.

Government policies may create inefficient incentives for conversion of prime and unique agricultural land which would not be converted in an idealized world with no distortions. In particular, zoning and development controls themselves may create pressure for development of better agricultural land, which may not exist in the absence of controls. Zoning or development control restrictions create a price differential between land which is approved for development and land which is not approved, and this premium may make it profitable to convert the wrong parcels of land.

Correction of such inefficiencies may require intervention by government agencies above the local level. However, the correct policy prescription in any particular case would require a considerable amount of information. A simple policy prohibiting the conversion of the best grades of agricultural land will *not* in general be appropriate. Again, a statement by the provincial government on the 'social' value of prime or unique agricultural land would make it much easier for municipalities to minimize the possible inefficiencies.

Many economists and legal experts have advocated the setting up of some form of 'development rights' market as a means of enacting development controls or zoning. This sort of policy would seem to be especially suitable for the problem we have been discussing. The urban government would determine how many acres of land will be approved for development and then would auction off the rights to develop. In the absence of other significant distortions, the owners of the less fertile land would bid more for the right to develop (since their profits from development would be higher), and so the efficient allocation (given the development controls) would result.

TORONTO-CENTRED REGION PLAN

Provincial involvement in comprehensive regional land use and development planning in Ontario became a matter of policy in 1966 with the announcement of the Design for Development program. The first product of the new provincial planning process was the plan for the Toronto-centred region (TCR), which was adopted as policy in 1971.

A review of the extent to which the provincial government implemented the TCR plan during the eight years following its adoption suggests that, contrary to the goals of the plan, the distribution of population and employment within the region at present is not very different from what it would have been in the absence of the provincial regional planning and development program. More important, little has been done which would suggest that these programs would have a significant effect on the distribution of economic activity in the year 2001. Few of the actions which would have been required to achieve the TCR goals were taken, and a number of actions which were taken conflicted with the goals.

Having reached this conclusion, however, it is important to add two points. First, the fact that the goals were not pursued relentlessly is not necessarily to be regretted, as two critical reviews of the TCR program have taken for granted. The case for the TCR program on economic efficiency grounds is open to question. Although many of the objectives were perhaps non-economic, the economic costs of the program were not seriously considered by either the politicians or the planners when the program was being adopted. Second, the fact that implementation of the TCR plan has been limited does not imply that the program had no effect whatsoever. A number of measures which were influenced by TCR plan considerations are documented in Chapter 7.

POLICIES OTHER THAN LAND USE CONTROLS

Although most public discussion of the existing and proposed role of the provincial government in ameliorating perceived land use problems has focused on provincial policies involving direct regulation of land use, we are convinced that other provincial policies often have a more important effect on land use in the province. One obvious example is the provincial responsibility in the provision of interurban servicing and transportation systems. The location, timing, and size of such systems have extremely important effects on the location, timing, and extent of development pressures. It was pointed out in Chapter 6, for example, that the decision to locate the Queen Elizabeth Way below the escarpment has probably contributed significantly to the pressures for development in the tender fruitlands. Furthermore, we argued in Chapter 4 that the provincial role in the provision of these services confers market power on the province, through its ability to control development. The province may exercise this power in order to confer benefits on one of its largest constituencies (homeowners).

The connection between the location of servicing and transportation systems and development pressures is certainly recognized by policy-makers. However, the inconsistencies which have sometimes arisen between provincial land use objectives and the development pressures resulting from the location of servicing and transportation systems make it appear that these conflicts have not been properly anticipated. These conflicts have also created uncertainty about the 'firmness' of provincial objectives, which has reduced the efficiency of the development process and of municipal policy-making.

Many provincial policies which do not affect land use directly have a very important *indirect* effect on the pattern and rate of change of land use. It is our opinion that such indirect policies in fact have a greater net effect on the pattern and rate of change of land use than any provincial policies concerned directly with affecting land use. One of the most important examples is provincial (and federal) direct and indirect subsidization of agriculture. The effect of these policies is to reduce the rate of conversion of agricultural land to non-agricultural uses. Such an effect may be socially desirable, but it must be recognized in any discussion of the desirability of enacting additional restrictions on the rate of conversion. When evaluating the efficiency of existing land use patterns and trends, the effects of such distortionary policies must not be ignored.

Bibliography

Acheson, K. (1977) 'Revenue vs. protection: the pricing of wine by the Liquor Control Board of Ontario.' *Canadian Journal of Economics* 10 (May), 246-62

Adler, G.M. (1971) *Land Planning by Administrative Regulation: The Policies of the Ontario Municipal Board* (Toronto: University of Toronto Press)

Alonso, W. (1960) 'A theory of the urban land market.' *Papers and Proceedings of the Regional Science Association* 6, 149-57

– (1964) *Location and Land Use* (Cambridge, Mass.: Harvard University Press)

Anderson, J.S. (1971) 'The relationship between soil class and forage yield.' M SC thesis, University of Guelph

Arnott, R. and J. Stiglitz (1975) 'Aggregate land rents, aggregate transport costs and expenditure on public goods.' Mimeographed. Institute for Economic Research, Queen's University, Kingston, Ontario

Arrow, K.J. (1963) 'Uncertainty and the welfare economics of medical care.' *American Economic Review* 53, 941-73

– (1964) 'The role of securities in the optimal allocation of risk-bearing.' *Review of Economic Studies* 31, 91-6

Arrow, K.J. and R.C. Lind (1970) 'Uncertainty and the evaluation of public investment decisions.' *American Economic Review* 60, 364-78

Avey, E.M. (1974) *Urban Spread into the Countryside North-west of London, Ontario* (London, Ont.: Department of Geography, University of Western Ontario)

Bacher, J. (1978) 'Fruitland battle continues.' *City Magazine* 3 (September), 17-19

Baumol, W.J. and W.E. Oates (1975) *The Theory of Environmental Policy* (Englewood Cliffs, NJ: Prentice-Hall)

Bens, C.K., A. Golden, and P. Bryant (1977) 'Design for Development: where are you?' Mimeographed. Bureau of Municipal Research, Toronto.

Bird and Hale, Ltd. and M.M. Dillon, Ltd. (1977) *Planning Goals - Food, Employment and Housing* (Toronto: Urban Development Institute)

Blumenfeld, H. (1978) Review of C. Beaubien and R. Tabacnik, *Perceptions 4: People and Agricultural Land* (Science Council of Canada 1977), in *Urban Forum* 3, 28-37

Boadway, R.W. (1974) 'The welfare foundations of cost-benefit analysis.' *Economic Journal* 84 (December), 926-39

Bossons, J. (1978) *Reforming Planning in Ontario: Strengthening the Municipal Role* (Toronto: Ontario Economic Council)

Bourne, L.S. (1975) *Urban Systems: Strategies for Regulation* (Oxford: Clarendon Press)

- (1977) 'The housing supply and price debate: divergent views and policy consequences.' Mimeographed. Centre for Urban and Community Studies, University of Toronto.

Breton, A. (1975) 'A positive theory of land use regulation.' Mimeographed. Institute for the Quantitative Analysis of Social and Economic Policy, University of Toronto.

Bryant, C.R. (1976) 'Some new perspectives on agricultural land use in the rural-urban fringe.' *Ontario Geography* 10, 64-78

Bryant, C.R. and L.H. Russwurm (1978) 'The impact of non-farm development on agriculture: a review of the arguments.' Mimeographed. Faculty of Environmental Studies, University of Waterloo.

Bureau of Municipal Research (1974) *The Development of New Communities in Ontario* (Toronto)

- (1977) *Food for the Cities: Disappearing Farmland and Provincial Land Policy* (Toronto)

Canada, Environment Canada, Lands Directorate (1977) *Land Use Programs in Canada: Ontario* (Ottawa)

Canada, Federal/Provincial Task Force on the Supply and Price of Serviced Residential Land (1978) *Down to Earth*, Vol. 1 (Ottawa)

County of Lambton, Planning Department (1973) *Consents in Lambton County*

Cullingworth, J.B. (1978) *Ontario Planning: Notes on the Comay Report on the Ontario Planning Act* (Toronto: Department of Urban and Regional Planning, University of Toronto)

Derkowski, A. (1972) *Residential Land Development in Ontario* (Urban Development Institute)

- (1975) *Costs in the Land Development Process* (Housing and Urban Development Association of Canada)

Dewees, D.N., C.K. Everson, and W.A. Sims (1975) *Economic Analysis of Environmental Policies* (Toronto: University of Toronto Press for Ontario Economic Council)
Dorfman, R., ed. (1965) *Measuring Benefits of Government Investments* (Washington, DC: The Brookings Institution)
Downing, P.B., ed. (1977) *Local Service Pricing Policies and Their Effect on Urban Spatial Structure* (Vancouver: University of British Columbia Press)
Ellickson, R. (1977) 'Suburban growth controls: an economic and urban analysis.' *Yale Law Journal* 388-511
Estrin, D. et al. (1978) *Environment on Trial: A Handbook of Ontario Environmental Law*, rev. ed. (Toronto: Canadian Environmental Law Research Foundation)
Flatters, F., V. Henderson, and P. Mieszkowski (1974) 'Public Goods, efficiency, and regional fiscal equalization.' *Journal of Public Economics* 3, 99-112
Frankena, M. (1978) 'An economic evaluation of urban transportation planning in Canada.' Mimeographed. Department of Economics, University of Western Ontario
– (1979) *Urban Transportation Economics: Theory and Canadian Policy* (Toronto: Butterworths)
Frankena, M. and D. Scheffman (1979) 'A theory of development controls in a 'small' city.' Mimeographed. Department of Economics, University of Western Ontario
Gertler, L.O. (1968) *Niagara Escarpment Study Fruit Belt Report* (Ontario Department of Treasury and Economics)
– (1972) *Regional Planning in Canada* (Montreal: Harvest House)
Gierman, D.M. (1976) *Rural Land Use Changes in the Ottawa-Hull Urban Region* (Ottawa: Environment Canada)
– (1977) *Rural to Urban Land Conversion* (Ottawa: Fisheries and Environment Canada)
Hamilton, B.W. (1978) 'Zoning and the exercise of monopoly power.' *Journal of Urban Economics* 5, 116-30
Henderson, J. (1977) *Economic Theory and the Cities* (New York: Academic Press)
Hirschleifer, J. (1976) *Price Theory and Applications* (Englewood Cliffs, NJ: Prentice-Hall)
Hoffman, D.W. (1971) *The Assessment of Soil Productivity for Agriculture* (ARDA)
Hoffman, D.W. and H.F. Noble (1975) *Acreages of Soil Capability Classes for Agriculture in Ontario* (Ontario Ministry of Agriculture and Food)

Horton, J.T. (1977) 'The Niagara Escarpment: planning for the multi-purpose development of a recreational resource.' In Krueger and Mitchell (1977), 149-72

Jackson, J.N. (1976) *Land Use Planning in the Niagara Region* (Toronto: Niagara Region Study Review Commission)

James F. MacLaren, Ltd. (1975) *Countryside Planning: A Methodology and Policies for Huron County and the Province of Ontario* (Toronto/London)

Kanemoto, Y. (1979) *Theories of Urban Externalities* (New York: Academic Press, forthcoming)

Krueger, R.R. (1978a) 'Methods of mapping degree of urbanization in the Niagara fruit belt.' Poster paper, University of Waterloo

– (1978b) 'Urbanization of the Niagara fruit belt.' *The Canadian Geographer* 22 (Fall), 179-94

Krueger, R.R. and R.C. Bryfogle, eds. (1971) *Urban Problems: A Canadian Reader* (Toronto: Holt, Rinehart and Winston)

Krueger, R.R. and B. Mitchell, eds. (1977) *Managing Canada's Renewable Resources* (Toronto: Methuen)

Krueger, R.R., et al., eds. (1970) *Regional and Resource Planning in Canada*, rev. ed. (Toronto: Holt, Rinehart and Winston)

Leith, J.C. (1976) *Exploitation of Ontario Mineral Resources: An Economic Policy Analysis* (Toronto: Ontario Economic Council)

Manne, H.G., ed. (1975) *The Economics of Legal Relationships* (St Paul: West Publishing Co.)

Margolis, J. (1956) 'On Municipal land policy for fiscal gains.' *National Tax Journal* 247-57

Markusen, J. and D. Scheffman (1977a) 'Ownership concentration in the urban land market: analytical foundations and empirical evidence.' In Smith and Walker (1977), 147-76

– (1977b) *Speculation and Monopoly in Urban Development: Analytical Foundations with Evidence for Toronto* (Toronto: University of Toronto Press for Ontario Economic Council)

– (1978a) 'Ownership concentration and market power in urban land markets.' *Review of Economic Studies* 45 (October), 519-26

– (1978b) 'The timing of residential land development: a general equilibrium approach.' *Journal of Urban Economics* 5 (October), 411-24

Martin, L.R.G. (1975) *Land Use Dynamics in the Toronto Urban Fringe* (Ottawa: Lands Directorate, Environment Canada, Information Canada)

Martin, L.R.G., and D.L. Matthews (1977) 'Recent land market activity on the Toronto rural-urban fringe.' *Urban Forum* 3, 18-25

Michie, G.H., and W.C. Found (1976) 'Rural estates in the Toronto region.' *Ontario Geography* 10, 15-26
Mills, E. and W. Oates, eds. (1975) *Fiscal Zoning and Land Use Controls* (Lexington, Mass.: Heath)
Mills, E.S. (1967) 'An aggregative model of resource allocation in a metropolitan area.' *American Economic Review* 57 (May), 197-210
Muller, R.A. (1978) *The Market for New Housing in the Metropolitan Toronto Area* (Toronto: Ontario Economic Council)
Muth, R.F. (1961) 'The spatial structure of the housing market.' *Papers and Proceedings of the Regional Science Association* 7, 207-20
- (1969) *Cities and Housing* (Chicago: University of Chicago Press)
Niagara Region Study Review Commission (1977) *Niagara Region: The Report of the Niagara Region Study Review Commission*
Ohls, J. and D. Pines (1975) 'Discontinuous urban development and economic efficiency.' *Land Economics* 224-34
Ontario, Advisory Task Force on Housing Policy (1973a) *Land Assembly and Servicing of Land* (Toronto)
- (1973b) *Land for Housing* (Toronto)
- (1973c) *Report* (Toronto)
Ontario, Central Ontario Lakeshore Urban Complex Task Force (1974) *Report to the Advisory Committee on Urban and Regional Planning*
Ontario, Commission of Inquiry into the Acquisition by the Ministry of Housing of Certain Lands in the Community of North Pickering (1978) *Report*
Ontario, Department of Municipal Affairs (annual 1961-71) *Annual Report*
Ontario, Department of Municipal Affairs, Metropolitan Toronto and Region Transportation Study (1968) *Choices for a Growing Region* (Toronto)
Ontario, Department of Planning and Development (1957) 'Manitouwadge.' *Ontario Planning* 4, No. 2, Supplement
Ontario, Department of Treasury and Economics, Regional Development Branch, Niagara Escarpment Study Group (1968) *Niagara Escarpment Study: Conservation and Recreation Report*
Ontario, Government of Ontario (1966) *Design for Development: Statement by the Prime Minister of the Province of Ontario on Regional Development Policy*
- (1968) *Design for Development: Phase Two*
- (1970a) *Presentation of Design for Development: Toronto Centred Region: Opening Remarks by Prime Minister of Ontario*
- (1970b) *Design for Development: The Toronto-Centred Region*
- (1971a) *Design for Development: A Policy Statement on the Northwestern Ontario Region*

- (1971b) *Design for Development: A Status Report on the Toronto-Centred Region*
- (1972) *Design for Development: Phase Three*
- (1978) *The Parkway Belt West Plan*

Ontario, Hearing Officers - Parkway Belt West (1977) *Report*

Ontario, Ministry of Agriculture and Food (annual) *Agricultural Statistics for Ontario*
- (1972) *Planning for Agriculture in Southern Ontario*, ARDA Report 7
- (1974) *Report of the Farm Classification Advisory Committee*
- (1976a) *A Strategy for Ontario Farmland*
- (1976b) *Agricultural Code of Practice*
- (1977a) *Green Paper on Planning for Agriculture: Food Land Guidelines*
- (1977b) *1976 Fruit Tree Census*

Ontario, Ministry of Housing (annual 1974/75 to 1977/78) *Annual Report*
- (1975) *North Pickering Project: Summary of Recommended Plan*
- (1976) *Land Severance*
- (1977a) *Highlights of the Townsend Community Plan*
- (1977b) *Ontario Housing Requirements 1976-2001* (Report and Appendices), prepared by Peter Barnard Associates
- (1977c) *Procedural Guidelines on Consent*
- (1977d) *Rural Estate Guidelines*
- (1978) *Land-Use Policy Near Airports*

Ontario, Ministry of Natural Resources (1976) *Ontario's Public Land*
- (1977) *Statistics*
- (1978) *1977 Annual Report*

Ontario, Ministry of Northern Affairs (1977) *Design for Development: Northwestern Ontario Initiatives and Achievements*

Ontario, Ministry of Treasury, Economics and Intergovernmental Affairs (formerly Department of Treasury and Economics) (annual 1970-77) *Budget*
- (1971) *Haldimand-Norfolk Study*, Vol. 1
- (1972) *Haldimand-Norfolk Study*, Vol. 2
- (1973a) *Development Planning in Ontario: The Niagara Escarpment*
- (1973b) *Development Planning in Ontario: The Parkway Belt: West*
- (1976) *Parkway Best West Draft Plan*
- (1977) *Recommended Parkway Belt West Development Plan*
- (1978) *Northwestern Ontario: A Strategy for Development*

Ontario, Ministry of Treasury, Economics and Intergovernmental Affairs, Regional Planning Branch (1976a) *Design for Development: Ontario's Future: Trends and Options*

– (1976b) *The Durham Subregion: A Strategy for Development to 1986*
– (1976c) *Toronto-Centred Region Program Statement*
Ontario, Niagara Escarpment Commission (1977) *Preliminary Proposals*
Ontario, Niagara Escarpment Task Force (1972) *To Save the Escarpment*
Ontario, Niagara Region Local Government Review Commission (1966), *Report*
Ontario, Northumberland Area Task Force (1975) *Final Report on the Preferred Development Strategy for the County of Northumberland*
Ontario, Ontario Economic Council (1973) *Subject to Approval: A Review of Municipal Planning in Ontario* (Toronto)
Ontario, Ontario Land Corporation (1976/77 and 1977/78) *Annual Report*
Ontario, Parkway Belt West Interested Groups and Residents Advisory Committee (1975) *Report*
Ontario, Planning Act Review Committee (1977) *Report*
Ontario, Royal Commission on Metropolitan Toronto (1975) *The Planning Process in Metropolitan Toronto*
– (1977a) *Local Decision-Making and Administration*
– (1977b) *Report*, Vols. 1 and 2
Ontario, Select Committee on the Ontario Municipal Board (1972) *Report*
Ontario, Simcoe-Georgian Area Task Force (1976) *Simcoe-Georgian Area Development Strategy Study*
Ontario, Special Committee on Farm Income in Ontario (1969) *The Challenge of Abundance*
Ontario, TCR Combined Task Forces and Regional Development Branch, Ministry of Treasury, Economics and Intergovernmental Affairs (1972) *Toronto-Centred Region Zone 1 and North-South Axis*, Vol. 1
Ontario Municipal Board Reports (quarterly) (Agincourt: Canada Law Book Limited)
Orr, D. (1976) *Property, Markets, and Government Intervention* (Pacific Palisades, Cal.: Goodyear)
Oxley, M.J. (1975) 'Economic theory and urban planning.' *Environment and Planning A* 7, 497–508
Peltzman, S. (1976) 'The regulation of automobile safety.' In H.L. Manne and R.L. Miller, eds., *Auto Safety Regulation: The Cure or the Problem?* (Glen Ridge, NJ: Horton) 1–52
Polinsky, A.M. (1978) 'On the choice between property rules and liability rules.' Mimeographed. Harvard Institute of Economic Research, Cambridge.
Pontryagin, L. et al. (1962) *The Mathematical Theory of Optimal Processes* (New York: Interscience)
Posner, R.A. (1971) 'Taxation by regulation.' *Bell Journal of Economics and Management Science* 2, 22–50

– (1974) 'Theories of economic regulation.' *Bell Journal of Economics and Management Science* 5, 335–58
– (1977) *Economic Analysis of Law*, 2nd ed. (Boston: Little, Brown)
Punter, J.V. (1974) *The Impact of Exurban Development on Land and Landscape in the Toronto-Centred Region: 1954–1971* (Ottawa: Central Housing and Mortgage Corporation)
Reeds, L.G. (1969) *Niagara Region Agricultural Research Report* (Regional Development Branch, Ontario Department of Treasury and Economics)
Regional Municipality of Niagara, Planning and Development Department (1976) *Review of the Regional Urban Areas Boundaries*. D.P.D. 750
– (1977a) *Population and Housing Growth Estimates for the Niagara Region*. D.P.D. 916
– (1977b) *Response to Urban Areas Boundaries Reduction*. D.P.D. 911 (revised)
– (1977c) *Residential Land Capacity and Housing Construction Estimates for the Niagara Region: 1976–96*. D.P.D. 997
– (1977d) *Rural Area Review: Report 3 – Development Patterns*
Richardson, H. (1973) *The Economics of Urban Size* (Lexington, Mass.: Lexington Books)
Robinson, I.M. (1977) 'Trends in provincial land planning, control, and management.' *Plan Canada* 17, 166–83
Rodd, R.S. (1975) *Use and Non-Use of Rural Land* (University of Guelph, Guelph)
– (1976a) 'The crisis of agricultural land in the Ontario countryside.' *Plan Canada* 16, 160–70
– (1976b) 'The use and abuse of rural land.' *Urban Forum* 2, 5–12
Rodd, R.S. and W. Van Vuuren (1975) 'A new methodology for countryside planning.' *Canadian Journal of Agricultural Economics* 23 (Workshop Proceedings), 109–40
Russwurm, L.H. (1967) 'Expanding urbanization and selected agricultural elements: cast study, southwestern Ontario area, 1941–1961.' *Land Economics* 43, 101–7
– (1970) *Development of an Urban Corridor System, Toronto to Stratford Area, 1941–1966* (Regional Development Branch, Ontario Department of Treasury and Economics)
– (1974) 'The countryside in Ontario: an overall policy and planning viewpoint.' In M.J. Troughton, J.G. Nelson, and S. Brown, eds. (1974) *The Countryside in Ontario* (London, Ont.: Department of Geography, University of Western Ontario)
– (1976) 'Country residential development and the regional city form in Canada.' *Ontario Geography* 10, 79–96

- (1977) *The Surroundings of Our Cities* (Ottawa: Community Planning Association of Canada)
Scheffman, D. (1977) 'Some evidence on the recent boom in house and land prices.' Forthcoming in L.S. Bourne and J.R. Hitchcock, eds. *Urban Housing Markets* (Toronto: University of Toronto Press)
Siegan, B.H. (1972) *Land Use without Zoning* (Lexington, Mass.: Heath)
Smith, L.B. and M. Walker, eds. (1977) *Public Property? The Habitat Debate Continued* (Vancouver: The Fraser Institute)
Solow, R. (1973) 'On equilibrium models of urban location.' In M. Parkin and A.R. Nobay, eds. *Essays in Modern Economics* (London: Longmans)
Statistics Canada, Census of Canada (quinquennial) *Agriculture*
Stigler, G. (1971) 'The theory of economic regulation.' *Bell Journal of Economics and Management Science* 2, 3-21
Stiglitz, J. (1977) 'The theory of local public goods.' In M. Feldstein and R. Inman, eds. *Economics of Public Services* (London: Macmillan)
Sugden, R. and A. Williams (1978) *The Principles of Practical Cost-Benefit Analysis* (Oxford: Oxford University Press)
Thoman, R.S. (1971) *Design for Development in Ontario: The Initiation of a Regional Planning Program* (Toronto: Allister Typesetting and Graphics)
Tiebout, C.M. (1956) 'A pure theory of local expenditures.' *Journal of Political Economy* 64, 416-24
Tindal, C.R. (1973) 'Regional Development in Ontario.' *Canadian Public Administration* 16, 110-23
Urban Development Institute, Region of York Committee (1972) *Brief on Design for Development* (Don Mills)
Vito, V. (1973) 'Erosion on the parkway belt?' (Toronto: Bureau of Municipal Research)
White, M. (1975) 'Fiscal zoning in fragmented metropolitan areas.' In Mills and Oates (1975), 31-100
Whitney, W. (1977) 'Inflation in urban land markets: a study of London, Ontario.' PHD dissertation, Department of Economics, University of Western Ontario (in progress)
Wingo, L. (1961a) *Transportation and Urban Land* (Washington, DC: Resources for the Future)
- (1961b) 'An economic model of the utilization of land for residential purposes.' *Papers and Proceedings of the Regional Science Association*, 7, 191-205
Wolfe, J.N. (1972) *Cost Benefit and Cost Effectiveness* (London: George Allen and Unwin)

Ontario Economic Council Research Studies

1 Economic Analysis of Environmental Policies
D.N. DEWEES, C.K. EVERSON, and W.A. SIMS

2 Property Crime in Canada: an econometric study
KENNETH L. AVIO and C. SCOTT CLARK

3 The Effects of Energy Price Changes on Commodity Prices, Interprovincial Trade, and Employment
J.R. MELVIN

4 Tariff and Science Policies: applications of a model of nationalism
D.J. DALY and S. GLOBERMAN

5 A Theory of the Expenditure Budgetary Process
D.G. HARTLE

6 Resources, Tariffs, and Trade: Ontario's stake
J.R. WILLIAMS

7 Transportation Rates and Economic Development in Northern Ontario
N.C. BONSOR

8 Government Support of Scientific Research and Development: an economic analysis
D.G. McFETRIDGE

9 Public and Private Pensions in Canada: an economic analysis
J.E. PESANDO and S.A. REA jr

10 Speculation and Monopoly in Urban Development: analytical foundations with evidence for Toronto
J.R. MARKUSEN and D.T. SCHEFFMAN

11 Day Care and Public Policy in Ontario
M. KRASHINSKY

12 Provincial Public Finance in Ontario: an empirical analysis of the last twenty-five years
D.K. FOOT

13 Extending Canadian Health Insurance: options for pharmacare and denticare
R.G. EVANS and M.F. WILLIAMSON

14 Measuring Health: lessons for Ontario
A.J. CULYER

15 Residential Property Tax Relief in Ontario
R.M. BIRD and N.E. SLACK

16 Fiscal Transfer Pricing in Multinational Corporations
G.F. MATHEWSON and G.D. QUIRIN

17 Housing Programs and Income Distribution in Ontario
GEORGE FALLIS

18 Economic Analysis of Provincial Land Use Policies in Ontario
MARK W. FRANKENA and DAVID T. SCHEFFMAN

19 Distribution of Income and Wealth: theory and evidence for Ontario
CHARLES M. BEACH, DAVID E. CARD, and FRANK FLATTERS

Lightning Source UK Ltd.
Milton Keynes UK
UKHW010000210722
406167UK00001B/247